6th Young Researcher Meeting – L'Aquila 2015

AA005656

Journal of Physics: Conference Series Volume 689

L'Aquila, Italy
12 - 14 October 2015

Editors:

Fabio Agostini
Claudia Antolini
Maria Bossa
Giordano Cattani
Stefano Dell'Oro
Marco Di Stefano
Marta D'Angelo
Giacomo Fragione
Marina Migliaccio
Lorenzo Pagnanini
Davide Pietrobon
Emanuela Puscedolu
Matteo Serra
Francesco Stellato

ISBN: 978-1-5108-2183-5
ISSN: 1742-6588

Printed from e-media with permission by:

Curran Associates, Inc.
57 Morehouse Lane
Red Hook, NY 12571

Printed by Curran Associates, Inc. (2016)

Institute of Physics
Dirac House, Temple Back
Bristol BS1 6BE UK

Phone: 44 1 17 929 7481
Fax: 44 1 17 920 0979

techtracking@iop.org

Additional copies of this publication are available from:

Curran Associates, Inc.
57 Morehouse Lane
Red Hook, NY 12571 USA
Phone: 845-758-0400
Fax: 845-758-2633
Email: curran@proceedings.com
Web: www.proceedings.com

Table of contents

Volume 689

2016

6th Young Researcher Meeting, L'Aquila 2015
12–14 October 2015, L'Aquila, Italy

Accepted papers received: 1 February 2016
Published online: 1 March 2016

Astrophysics

Detectors for Particle Physics

Condensed Matter Physics

Physics of the Biological Systems

Journal of Physics: Conference Series **689** (2016) 011001 doi:10.1088/1742-6596/689/1/011001

Young Researcher Meeting, L'Aquila 2015

F Agostini, C Antolini, M Bossa, G Cattani, S Dell'Oro, M D'Angelo, M Di Stefano, G Fragione, M Migliaccio, L Pagnanini, D Pietrobon, E Pusceddu, M Serra, F Stellato

PREFACE

The Young Researcher Meeting (www.yrmr.it) has been established as a forum for students, postdoctoral fellows and young researchers determined to play a proactive role in the scientific progress. Since 2009, we run itinerant, yearly meetings to discuss the most recent developments and achievements in physics, as we are firmly convinced that sharing expertise and experience is the foundation of the research activity. One of the main purposes of the conference is actually to create an international network of young researchers, both experimentalists and theorists, and fruitful collaborations across the different branches of physics.

The format we chose is an informal meeting primarily aimed at students and postdoctoral researchers at the beginning of their scientific career, who are encouraged to present their work in brief presentations able to provide genuine engagement of the audience and cross-pollination of ideas.

The sixth edition of the Young Researcher Meeting was held at the Gran Sasso Science Institute (GSSI), L'Aquila. The high number of valuable contributions gave rise to a busy program for a two-day conference on October 12th-13th. The event gathered 70 participants from institutes all around the world. The plenary talk sessions covered several areas of pure and applied physics, and they were complemented by an extremely rich and interactive poster session.

This year's edition of the meeting also featured a *lectio magistralis* by Prof. E. Coccia, director of the GSSI, who discussed the frontiers in gravitational wave physics, commemorating the International Year of Light on the centenary of Einstein's theory of general relativity.

On October 14th, the participants to the conference took part to a guided tour of the Gran Sasso National Laboratories (LNGS), one of the major particle physics laboratories in the world.

In this volume, we collect part of the contributions that have been presented at the conference as either talks or posters. They cover topics in astrophysics and cosmology, particle and theoretical physics, soft and condensed matter, biophysics, medical physics, and instrumentation. Advancements in theoretical, experimental and data analysis fields are well balanced across the contributions. Marking the year of the Nobel Prize *for the discovery of neutrino oscillations, which shows that neutrinos have masses*, several contributions detail the recent results from ANTARES, an experiment optimised for the detection of muons from high-energy astrophysical neutrinos. As the meeting was held at the GSSI, a number of papers focus on the cutting-edge experimental developments for the experiments at the LNGS (e.g. LUNA, XENON). Moreover, interesting updates have been delivered by young researchers involved in experiments at other major facilities in the world, such as the Elettra Synchrotron, Sieroszowice Underground Laboratory, Fermilab and Jefferson Laboratory.

Journal of Physics: Conference Series **689** (2016) 011001 doi:10.1088/1742-6596/689/1/011001

YRM Organising and Editorial Committee

Fabio Agostini
(fabioagostini31@gmail.com)
NAIS, Italy

Claudia Antolini
(cla.antolini@gmail.com)

Maria Bossa
(maria.bossa@gssi.infn.it)
Gran Sasso Science Institute

Giordano Cattani
(giordano.cattani@gmail.com)

Stefano Dell'Oro
(stefano.delloro@gssi.infn.it)
Gran Sasso Science Institute

Marco Di Stefano
(marco.distefano@cnag.crg.eu) Centre Nacional d'Anàlisi Genòmica - Centre de Regulació Genòmica (CNAG-CRG)

Marta D'Angelo
(marta.dangelo@gssi.infn.it)
Gran Sasso Science Institute

Giacomo Fragione
(giacomo.fragione90@gmail.com)
Physics Department, "La Sapienza" University of Rome

Marina Migliaccio
(mm858@ast.cam.ac.uk)
Institute of Astronomy and Kavli Institute for Cosmology, University of Cambridge

Lorenzo Pagnanini
(lorenzo.pagnanini@gssi.infn.it)
Gran Sasso Science Institute

Davide Pietrobon
(daddeptr@gmail.com)
HERE, California

Emanuela Pusceddu
(emanuela.pusceddu@gmail.com)
Institute of Biometeorology CNR

Matteo Serra
(serra@mat.uniroma1.it)
Mathematics Department, "La Sapienza" University of Rome

Francesco Stellato
(francesco.stellato@roma2.infn.it)
INFN Roma Tor Vergata

ACKNOWLEDGMENTS

The organisers of the 6th Young Researcher Meeting, held in L'Aquila, would like to thank all the scientists who participated in the meeting.

We thank the Gran Sasso Science Institute (GSSI) and Laboratori Nazionali del Gran Sasso (LNGS) for hosting the conference, covering the organizational costs and the expenses for the publication of the proceedings, and providing travel grants to the speakers. We are grateful to Prof. Eugenio Coccia, director of GSSI, Prof. Francesco Vissani, coordinator of the Astroparticle Physics Ph.D. course at GSSI, Prof. Stefano Ragazzi, director of LNGS, and Dr. Alba Formicola, head of LNGS research division, for their invaluable support and advice.

The event was broadcast live on the GSSI website.

GRAN SASSO SCIENCE INSTITUTE
CENTER FOR ADVANCED STUDIES
INFN

INFN
Laboratori Nazionali del Gran Sasso

Journal of Physics: Conference Series **689** (2016) 011002 doi:10.1088/1742-6596/689/1/011002

Peer review statement

All papers published in this volume of *Journal of Physics: Conference Series* have been peer reviewed through processes administered by the proceedings Editors. Reviews were conducted by expert referees to the professional and scientific standards expected of a proceedings journal published by IOP Publishing.

Dark Matter and Baryon Asymmetry production from out-of-equilibrium decays of Supersymmetric states

Giorgio Arcadi[1]

[1]Laboratoire de Physique Théorique, CNRS, Btiment 210, Univ. Paris-Sud 11, 91405 Orsay Cedex, France

E-mail: `giorgio.arcadi@th.u-psud.fr`

Abstract. We will review the main aspects of a mechanism for the contemporary generation of the baryon and Dark Matter abundances from the out-of-equilibrium decay of a WIMP-like mother particle and briefly discuss a concrete realization in a Supersymmetric scenario.

1. Introduction

Two pressing puzzles of modern particle physics are the generation of the dark and visible matter components of the Universe. As further confirmed by high precision measurements by the Planck satellite [1], approximately one third of the total energy budget of the Universe is provided by a Dark Matter (DM) component, possibly represented by a cosmologically stable particle state with suppressed interactions with the visible matter. Given the absence of detection of DM, its nature and properties as well as its production mechanism are still unknown. The origin of the visible component of the Universe, contributing only to 5% of the total energy budget, is similarly puzzling. Indeed, the lack of observation of primordial antimatter needs a dynamical mechanism (baryogenesis) which has originated, at some stage of the history of the Universe, an asymmetry between matter and antimatter. Although the exact mechanism is still unknown, there are three necessary conditions which should be satisfied [2]: violation of the baryon (B) number, violation of CP, departure from thermal equilibrium.

The solution of these two puzzles requires an extension of the Standard Model of Particle Physics (SM), since it does not provide for a viable DM candidate and does not fulfill in a satisfactory way the necessary conditions for baryogenesis. A very simple possibility is to generate the baryon asymmetry and the DM abundance from the out-of-equilibrium decay of an exotic state [3, 4] (see also [5–8] for similar ideas).

We will provide, in this work, a pedagogical and possibly not too technical review of the general aspects of this mechanism specializing, in the last section, on the specific example of a realization of this kind of scenario in the Minimal Supersymmetric Standard Model (MSSM).

2. General idea

The mechanism discussed in this work relies on the idea of generating both the DM and baryon abundances from the out-of-equilibrium decay of a particle state X featuring B-number and CP violating interactions. The two abundances can be schematically expressed as:

$$\Omega_{\Delta B} = \epsilon_{\rm CP}\frac{m_p}{m_X}\, BR(X \rightarrow b, \bar{b})\, \Omega_X, \tag{1}$$

$$\Omega_{DM} = \frac{m_{DM}}{m_X} BR\,(X \rightarrow DM + \text{anything})\, \Omega_X \tag{2}$$

where we have used, as customary, the adimensional parameters $\Omega_i \equiv \frac{\rho_i}{\rho_c}$ with ρ_i and ρ_c being, respectively, the energy density of the considered species and the so-called critical density.

The two expressions depend on the initial abundance $\Omega_{\rm X}$ of the mother particle, weighted by the branching fraction in the considered channel (in the first expression $b, \bar{b}$ refer, respectively, to any combination of SM states with non zero baryon number and its charge conjugate; the decay into DM is assumed in general to be B-preserving), and the ratio between the mass of the considered species (in the case of the baryons there is the mass of the proton m_p) and the decaying particle. The baryon abundance depends as well on the CP-asymmetry $\epsilon_{\rm CP}$, defined as:

$$\epsilon_{\rm CP} = \frac{\Gamma(X \rightarrow b) - \Gamma(X \rightarrow \bar{b})}{\Gamma(X \rightarrow b) + \Gamma(X \rightarrow \bar{b})}\,. \tag{3}$$

From the expressions above it is straightforward to evince that the correct amount of baryons and DM can be generated from a suitable initial abundance of the mother particle. In this work we will consider the case that this is a WIMP (Weakly Interacting Massive Particle)-like particle. By this we mean that the mother particle existed in thermal equilibrium at early stages of the history of the Universe after which it has undergone chemical freeze-out. The time of freeze-out and the corresponding abundance are determined by the pair annihilation rate of this particle. For conventional WIMPs the freeze-out occurs in a strongly non-relativistic regime, i.e. temperatures of the Early Universe sensitively lower than the mass of the particle, typically $T_{\rm f.o.} = m_X/20 - m_X/30$, and the relic density Ω_X is inversely proportional to the thermally averaged pair annihilation rate $\langle \sigma v \rangle$. Contrary to conventional WIMPs, the particle X is, however, metastable and decays after freeze-out producing the baryon asymmetry and the DM abundance. The most relevant feature of this framework is that the relevant quantities are substantially determined by particle physics inputs. As a consequence the new particles and interactions needed for the generation of the baryon and DM abundances can be potentially probed by Earth scale experiments. Although the expressions (1)-(2) provide a very good description of the main features of the mechanism and can provide

Journal of Physics: Conference Series **689** (2016) 012001 doi:10.1088/1742-6596/689/1/012001

a good estimate of the considered quantities in simple models, they should be replaced by a more refined treatment once realistic particle physics frameworks are considered. This refinement consists in the numerical solution of a system of Boltzmann equations. We will briefly illustrate in the next section the main features of this kind of system and how to quantitatively compute the DM and baryon abundances.

3. Quantitative determination of the Baryon and DM abundances

The simple picture depicted in the previous section is complicated by the following aspects: 1) the standard approximations for the computation of Ω_X according the WIMP paradigm are problematic in the scenario under consideration. Indeed, in most particle frameworks, $\epsilon_{\rm CP}$ and $BR(X \to DM + \text{anything})$ are very small numbers. They should be then compensated by a very high Ω_X which can be typically achieved considering a thermal freeze-out close to the relativistic regime. In addition, being the X state not stable, interaction vertices between single X states and SM particles are in general present, leading to single-annihilation processes in addition to the conventional pair annihilations, modifying the usual relation $\Omega_X \propto 1/\langle \sigma v \rangle$. 2) Processes generating baryon asymmetry are typically accompanied by the so called wash-out processes. These processes also violate the baryon number and contrast the former ones tending to restore a null net baryon number. The baryogenesis mechanism should be enough efficient to compensate their effect. 3) In order to have an efficient production of the baryon asymmetry the decaying state should typically feature both sizable B-conserving and B-violating decay channels [6, 9]. The simplest way to achieve this result is to introduce a third state ψ lighter than the mother one, but not being the DM itself [1], which triggers the B-conserving decays. At the same time the introduction of the new particle implies the presence of additional processes. In particular it takes a very relevant role in wash-out effects. 4) The baryon asymmetry can be as well produced in B-violating annihilations of the mother particle. 5) The decay of the mother particle might be not the only source of DM.

All these effects can be consistently treated by introducing three sets of coupled Boltzmann equations. The first traces the abundance of the X and ψ particles. The corresponding equations are schematically written as:

$$
\begin{aligned}
\frac{dY_{\tilde{\alpha}}}{dx} &= -\frac{1}{Hx}\Gamma_{\tilde{\alpha},\Delta B\neq 0}\left(Y_{\tilde{\alpha}} - Y^{\rm eq}_{\tilde{\alpha}}\right) - \frac{s}{Hx}\langle \sigma v\rangle_{\tilde{\alpha},\Delta B \neq 0} Y_X^{\rm eq}\left(Y_{\tilde{\alpha}} - Y^{\rm eq}_{\tilde{\alpha}}\right) \\
&- \frac{s}{Hx}\sum_{\tilde{\beta}\neq\tilde{\alpha}}\langle \sigma v\rangle \left(\tilde{\alpha}\tilde{\beta} \to X\right)\left(Y_{\tilde{\alpha}}Y_{\tilde{\beta}} - Y^{\rm eq}_{\tilde{\alpha}}Y^{\rm eq}_{\tilde{\beta}}\right) - \frac{s}{Hx}\sum_{\tilde{\beta}\neq\tilde{\alpha}}\langle \sigma v\rangle\left(\tilde{\alpha}X \to \tilde{\beta}X\right)Y_X^{\rm eq}\left(Y_{\tilde{\alpha}} - \frac{Y^{\rm eq}_{\tilde{\alpha}}}{Y^{\rm eq}_{\tilde{\beta}}}Y_{\tilde{\beta}}\right) \\
&- 2\frac{s}{Hx}\langle \sigma v\rangle_{\tilde{\alpha}\tilde{\alpha}}\left(Y^2_{\tilde{\alpha}} - Y^{\rm eq\,2}_{\tilde{\alpha}}\right) - \frac{1}{Hx}\sum_{\tilde{\beta}\neq\tilde{\alpha}}\Gamma_{\Delta B=0}\left(Y_{\tilde{\alpha}} - Y^{\rm eq}_{\tilde{\alpha}}\frac{Y_{\tilde{\beta}}}{Y^{\rm eq}_{\tilde{\beta}}}\right) - \frac{1}{Hx}\Gamma\left(\tilde{\alpha} \to DM\right)Y_{\tilde{\alpha}}, \\
&\tilde{\alpha}, \tilde{\beta} = X, \psi
\end{aligned}
\tag{4}
$$

where we have introduced the adimensional parameter $Y_i = n_i/s$ (n_i is the number density of the considered quantity while s is the entropy density) with the ratio

[1] Possible B-conserving decays into DM cannot provide the desired effect because of the typically very low branching fraction.

$x = m_X/T$ as independent variable. $H \propto \frac{m_X^2}{x^2 M_{\rm Planck}}$ is the Hubble parameter, assuming standard cosmological history. The first line represents the processes originating the baryon asymmetry (it is zero for ψ). In general the baryon asymmetry can also be created by B and CP violating $2 \to 2$ scatterings. The corresponding contribution is however negligible in the specific example proposed (see instead [10–12], where some cases in which the baryon asymmetry is mostly produced by scatterings). The second line and the first term in the third line represent the other annihilation processes, respectively with $X\psi$ as initial state (coannihilations), single B-conserving annihilations $X + SM \to \psi + SM$ and conventional pair annihilations. The last term is the production of DM. In concrete realizations there might be more than two particles contributing to the relevant processes. In such a case there will be a larger set of equations with the same general structure depicted above. We have then the equation for the baryon asymmetry, customarily casted in terms of $Y_{\Delta B-L}$, with L being the lepton number, in order to get rid of the sphaleron processes, since they conserve $B - L$:

$$\frac{dY_{\Delta B-L}}{dx} = \frac{1}{Hx}\Delta\Gamma_{X,\Delta B\neq 0}\left(Y_X - Y_X^{\rm eq}\right) + \frac{s}{Hx}\langle \Delta\sigma v\rangle_X \left(Y_X - \frac{Y_X^{\rm eq}}{Y_\psi^{\rm eq}} Y_\psi\right) - {\rm WO}(\mu_{\rm f}). \quad (5)$$

Here the first line represents the sources of baryon asymmetry, with $\Delta\Gamma_{X,\Delta B} = \epsilon_{\rm CP}\Gamma_{X,{\rm tot}}$. The term in the second line, schematically labeled as WO (the full description is given in [4]), represents the ensemble of wash-out processes, which, whether efficient at the time or after the generation of the baryon asymmetry, deplete the baryon abundance. The wash-out terms are proportional to the chemical potentials $\mu_f \propto n_f - n_{\bar f}$, where $f =$ SM quark. The last and simplest equation is the one for the DM abundance:

$$\frac{dY_{DM}}{dx} = \frac{1}{Hx}\sum_{\tilde\alpha}\Gamma\left(\tilde\alpha \to DM\right). \quad (6)$$

In general the DM is the lightest stable state of a new particle sector. As a consequence all the states belonging to this sector will decay into it ($\tilde\alpha$ then runs over all the states of the theory). This can lead to an additional production mechanism of DM dubbed freeze-in (a complete list of references is provided in [4]) consisting in the production from states still in thermal equilibrium.
This system is solved by setting the initial condition, at $x \ll 1$, $Y_{DM} = Y_{\Delta {\rm B-L}} = 0$ and the abundances of X and ψ fixed to the equilibrium value, and by setting a final time x_f corresponding to $x \gg x_d$, with x_d being the decay time of X. Eq. (1)-(2) are related to the solution of the Boltzmann system by $\Omega_{\Delta B,{\rm DM}} \propto m_{\rm p,DM} Y_{\Delta B,DM}(x_f), Y_{\Delta B} = \frac{28}{79} Y_{\Delta B-L}$.
In the most minimal realization there are 5 relevant parameters: the three masses $m_{\rm DM}, m_X, m_\psi$, the coupling λ associated to the B-violating processes, and the scale M which characterizes the size of the annihilation and decay processes of the X, ψ particles and which can be identified as the mass of a fourth category of new particles.
In the next section we will illustrate some results in a specific example.

4. Concrete example

The general framework depicted above is straightforwardly realized in the MSSM. Indeed its rich particle spectrum provide the new particles useful for the considered mechanism.

Journal of Physics: Conference Series **689** (2016) 012001 doi:10.1088/1742-6596/689/1/012001

Moreover baryon number is in general violated in the MSSM unless one imposes a discrete symmetry dubbed R-parity. In absence of R-parity symmetry, the superpartners of the standard baryons and leptons decay into SM states. For example the operator $\lambda U^c D^c D^c$ is responsible for the baryon number violation, preserving the lepton number, avoiding conflicts with the stability of the proton. The DM candidate is the Gravitino which is not exactly stable but has a lifetime exceeding by orders of magnitude the lifetime of the Universe thanks to the Planck scale suppression of its interactions [2]. In addition, more sources of CP violation are present in the MSSM. The role of the X state is played by the Bino while ψ is one the two other gauginos. In the considered example ψ is the Gluino. The role of M is played by two energy scales, m_0 and μ, representing the mass scales of the scalar superpartners and of Higgsinos. The Bino features R-parity violating channels into SM fermions as well as R-parity conserving decays into the Gravitino DM and the Gluino. The CP asymmetry depends on the fundamental mass scales of the theory as $\epsilon_{\rm CP} \propto \frac{m_{\tilde{B}} m_{\tilde{G}}}{m_0^2} f\left(m_{\tilde{G}}^2/m_{\tilde{B}}^2\right)$ with the function f being 0 for $m_{\tilde{G}} \geq m_{\tilde{B}}$ and 1 for $m_{\tilde{G}} \ll m_{\tilde{B}}$.

The Bino-like mother particle has annihilation cross-sections going like $\langle \sigma v \rangle \propto \frac{m_{\tilde{B}}^2}{m_0^4}, \frac{1}{\mu^2}$ and a total decay rate $\Gamma_{\tilde{B}} \propto \frac{m_{\tilde{B}}^5}{m_0^4}$ (the detailed expressions can be found in [4]). The desired value of the high abundance and long lifetime are achieved for $m_0, \mu \gtrsim 10^6$ GeV resembling the so-called split and mini-split SUSY scenarios which have become rather popular on recent times since they can account the lack of detection of supersymmetric states at the LHC.

For the details of the solution of the Boltzmann equations and the determination of the viable parameter space of this scenario we refer to the original reference [4]. We just report in fig. (1) a sample plot of the solution of the Boltzmann equations for a generic assignation of parameters.

Here have been reported the evolution of the abundaces of the three relevant species, i.e. the Bino (mother particle), the baryon asymmetry and the gravitino DM. The purple and blue dashed lines represent the simple scenario described at the beginning, in which the two abundances are contemporary generated by the Bino which occurs in a semi-relativistic freeze-out at $x \sim 2$ but decays at much later times. We have shown as well a possible complication due to the rich structure of Supersymmetric theory. We have indeed considered a parameter assignation in which a sizable DM component is generated by another particle of the spectrum, the Wino, heavier than the mother particle. It does not contribute to the baryogenesis process since its annihilation rate is typically very high and then its abundance after freeze-out is negligible. It can however efficiently generate the DM through the so called freeze-in mechanism (also the Bino is capable of this kind of contribution, but in this case it is negligible with respect to the out-of-equilibrium decay).

[2] In R-parity violating scenarios the Superpartners have decay chains which do not necessarily end into DM. The still feature branching fractions of decay into gravitino such that the considerations of the previous sections still apply.

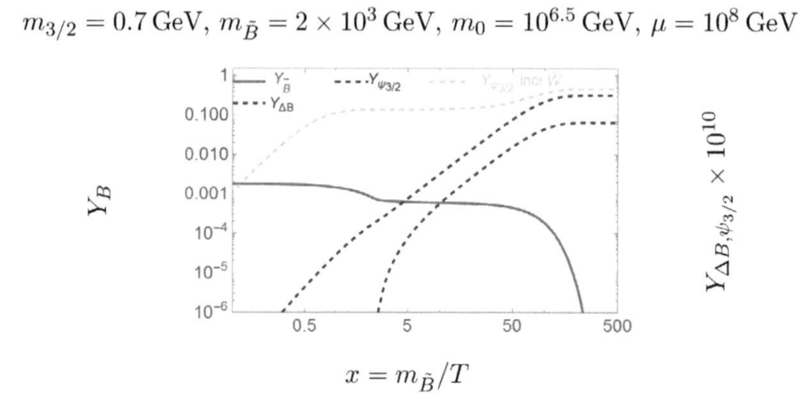

Figure 1: Sample plot showing the evolution of the abundances of the Bino ($Y_{\tilde{B}}$) (red solid curve), the baryons ($Y_{\Delta B}$) (purple dashed curve) and of the gravitino ($Y_{\psi_{3/2}}$) (blue and cyan curves). In the cyan curve is considered an additional sizable DM production from another Supersymmetric state, i.e the Wino. The assignation of the fundamental parameters is reported in the plot.

5. Conclusions

We have briefly reviewed the realization of a mechanism for the contemporary generation of the DM abundance and of the baryon asymmetry from the decay of a WIMP-like particle. This mechanism is simple and allow the prediction of the relevant quantities essentially in terms of a particle physics input. The concrete implementation in a definite particle physics framework requires a detailed numerical treatment based on the solution of suitable Boltzmann equations. We have shown an explicit implementation in a MSSM realization.

Acknowledgments: The author warmly thanks the organizers of the 6th Young Reseacher meeting for the chance of giving this contribution and the GSSI Institute for the hospitality. The author is supported by the ERC advanced Grant Higgs@LHC.

References

[1] Ade P A R *et al.* (Planck) 2015 (*Preprint* 1502.01589)
[2] Sakharov A 1991 *Sov.Phys.Usp.* **34** 417–421
[3] Arcadi G, Covi L and Nardecchia M 2014 *Phys. Rev.* **D89** 095020 (*Preprint* 1312.5703)
[4] Arcadi G, Covi L and Nardecchia M 2015 (*Preprint* 1507.05584)
[5] Cui Y and Sundrum R 2013 *Phys. Rev.* **D87** 116013 (*Preprint* 1212.2973)
[6] Rompineve F 2014 *JHEP* **08** 014 (*Preprint* 1310.0840)
[7] Cui Y 2013 *JHEP* **12** 067 (*Preprint* 1309.2952)
[8] Cui Y 2015 (*Preprint* 1510.04298)
[9] Nanopoulos D V and Weinberg S 1979 *Phys.Rev.* **D20** 2484
[10] Claudson M, Hall L J and Hinchliffe I 1984 *Nucl.Phys.* **B241** 309
[11] Baldes I, Bell N F, Petraki K and Volkas R R 2014 (*Preprint* 1407.4566)
[12] Baldes I, Bell N F, Millar A, Petraki K and Volkas R R 2014 *JCAP* **1411** 041 (*Preprint* 1410.0108)

Journal of Physics: Conference Series **689** (2016) 012002 doi:10.1088/1742-6596/689/1/012002

Study of a Scalar Field on the Maximally Extended Schwarzschild Spacetime

C Buss[1], **M Casals**[1,2]

[1]Centro Brasileiro de Pesquisas Físicas, Rua Dr. Xavier Sigaud, 150 - Urca Rio de Janeiro, Brasil

[2]School of Mathematics & Statistics and Complex & Adaptive Systems Laboratory, University College Dublin, Belfield, Dublin 4, Ireland

E-mail: cbuss@cbpf.br

Abstract. In this paper we present preliminary results about the investigation of scalar perturbations both inside and outside a Schwarzschild black hole. We express the wave equation in Ingoing Eddington-Finkelstein coordinates (explicitly showing its separability), write local series expansions for mode solutions and use them to then numerically calculate scalar perturbations inside and outside the future event horizon. Our main motivation is to calculate, in the future, the two-point quantum correlator in Schwarzschild spacetime, with points on opposite sides of the horizon.

1. Introduction

The study of linear field perturbations of black hole spacetimes is important for understanding the properties of black holes, both classical and quantum. Most studies of black hole spacetimes have concentrated on the exterior region of the black hole (see, e.g., [1–6]). Indeed, standard coordinate systems usually only cover the exterior region. There exist, however, coordinate systems which cover both the exterior and the interior regions of the black hole. Such systems are naturally suited for studying field perturbations propagating across the event horizon of the black hole. Investigating field perturbations inside the horizon is useful for various purposes. For example, classically, it has been shown that field perturbations across the event horizon lead to instabilities in the interior of certain black holes (see, e.g., [7–9]). Quantum-mechanically, the study of objects such as the two-point quantum correlator, with one point inside the horizon and the other point outside, may yield an insight into properties of Hawking radiation. This is precisely the main motivation on the base of our work.

Some studies of the quantum correlator have already been carried out in toy-model black hole spacetimes (e.g., [10, 11]). Our aim is the calculation of the quantum correlator for a scalar field in a non-rotating (Schwarzschild) black hole spacetime, with points on opposite sides of the horizon. With this aim in mind, in this paper we present preliminary results for the calculation of scalar field perturbations outside as well as *inside* a Schwarzschild black hole. For this purpose, we have chosen to work with Ingoing Eddington-Finkelstein (IEF) coordinates as opposed to the more widely used Schwarzschild (Schw) coordinates. The reason being that IEF coordinates not only render the wave field equation separable - as in Schwarzschild coordinates - but also they are regular across the (future) event horizon - as opposed to Schwarzschild coordinates.

Separability is a desirable property of the field equation as it faciliates the practical calculation of its solutions.

This paper is organized as follows. In Sec. 2 we introduce the maximal extension of Schwarzschild spacetime, Schwarzschild and IEF coordinate systems, the wave equation for a massles scalar field, and define various mode solutions. In Sec. 3 we present analytical results: we explicitly write out and separate the wave equation in IEF coordinates and we give series expansions for mode solutions. In Sec. 4 we present a numerical calculation of mode solutions both outside and inside the future event horizon of a Schwarzschild black hole.

2. Theoretical Background

Schwarzschild spacetime describes the (unique) spherically symmetric, asymptotically flat vacuum black hole solution of Einstein equations. These properties are manifest in the metric written in Schwarzschild coordinates:

$$ds^2 = -\left(1 - \frac{2M}{r}\right) dt^2 + \left(1 - \frac{2M}{r}\right)^{-1} dr^2 + r^2(d\theta^2 + \sin^2\theta d\phi^2). \tag{1}$$

Here, M is the mass of the black hole; r, outside the event horizon, is the radial coordinate associated with the circumference; t is the time as measured by a stationary observer at radial infinity; and θ and ϕ are, respectively, polar and azimuthal angles. The event horizon is located at $r = 2M$; we note that we use geometrized units ($G = 1$ and $c = 1$). These coordinates manifestly become irregular at the horizon. It is known, however, that this spacetime can be maximally extended, that is, geodesics in the maximally extended spacetime either start/end at a curvature singularity or are "continuable infinitely with respect to its 'natural length'" [13]. We show the maximal extension of Schwarzschild spacetime in the Penrose-Carter diagram of Fig. 1.

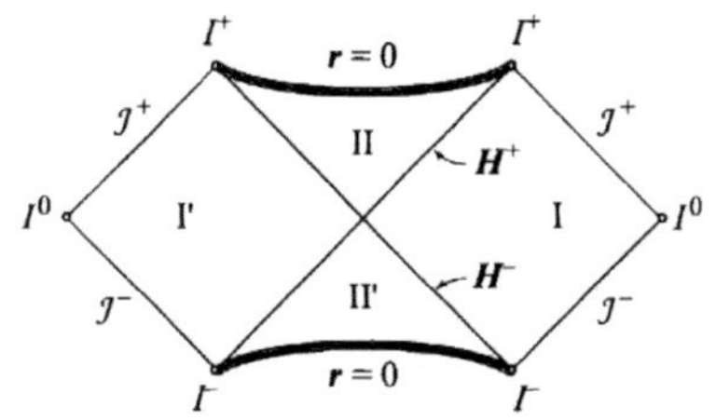

Figure 1. Penrose-Carter diagram showing the maximal extension of the Schwarzschild manifold. Region II is the black hole interior, II' is the white whole interior, while I and I' are the external regions. $\mathcal{H}^+$ and $\mathcal{H}^-$ are the future and past event horizons, respectively. $\mathcal{J}^+$ and $\mathcal{J}^-$, future and past null infinities. I^+ and I^- are the future and past timelike infinities, while I^0 are the spacelike infinities. Finally, $r = 0$ represent the curvature singularities. A massive particle in the neighborhood of an astrophysical Schwarzschild blackhole would typically start its trajectory at region I^- and either end up at I^+ without entering the black hole or else cross, perhaps, the future event horizon $\mathcal{H}^+$ entering region II. See [15].

In this work, we will use IEF coordinates. In this coordinate system, the metric is written as

$$ds^2 = -\left(1 - \frac{2M}{r}\right) dv^2 + 2dvdr + r^2(d\theta^2 + \sin^2\theta d\phi^2), \tag{2}$$

where t, θ and ϕ are as in Schwarzschild coordinates, while v is an advanced-time coordinate, and is related to Schwarzschild coordinates by $v = t + r_*$, with $r_* = r + \log\left|\frac{r-2M}{2M}\right| \in (-\infty, +\infty)$.

We note that Schwarzschild coordinates are irregular at $r = 2M$, i.e., both at $\mathcal{H}^+$ and at $\mathcal{H}^-$. On the other hand, the IEF coordinates are regular on the future event horizon $\mathcal{H}^+$, which corresponds to the horizon of an astrophysical black hole, as well as throughout regions I and II of the Penrose-Carter diagram. In fact, it is apparent from the equations for radial null geodesics, $dv/dr = 0$ and $dv/dr = 2(1-2M/r)^{-1}$, that this coordinate system describes well the infall of massless particles across $\mathcal{H}^+$.

In this work, we consider a massless scalar field perturbation of Schwarzschild spacetime. This corresponds to a fixed background metric supporting a scalar field φ obeying the following (Klein-Gordon) wave equation:

$$\nabla_\mu \nabla^\mu \varphi = 0. \tag{3}$$

In Schwarzschild coordinates, this wave equation is separable, allowing for the decomposition of the field into mode solutions, i.e., solutions depending on the separation constants as

$$\varphi(t, r, \theta, \phi) = \frac{e^{-i\omega t}}{r} R_{\omega l}(r) Y_{lm}(\theta, \phi), \tag{4}$$

where ω, l and m are the aforementioned separation constants, Y_{lm} is a spherical harmonic and $R_{\omega l}$ satisfies a linear, second order (radial) ordinary differential equation.

One can find a complete basis of mode solutions of the wave equation by choosing any two types of radial solutions defined by the boundary conditions given in Tables 1 and 2: In, Out and Up. It is of central importance for our work the results obtained by Damour and Ruffini [17], which give a prescription for the extension of the boundary conditions for the modes In and Out across the future event horizon. This is given in Table 2.

Table 1. Modes and Boundary Conditions - Region I.

Mode	Boundary Conditions		Mode	Boundary Conditions	
$R^{in}_{\omega l}(r) \sim$	$e^{-i\omega r_*}$,	$r_* \to -\infty$	$R^{up}_{\omega l}(r) \sim$	$B_{out} e^{i\omega r_*} + B_{in} e^{-i\omega r_*}$,	$r_* \to -\infty$
	$A_{out} e^{i\omega r_*} + A_{in} e^{-i\omega r_*}$,	$r_* \to \infty$		$e^{i\omega r_*}$,	$r_* \to \infty$

Table 2. Modes and Boundary Conditions - Regions I and II.

Mode	Boundary Conditions		Mode	Boundary Conditions	
$R^{in}_{\omega l}(r) \sim$	$e^{-i\omega r_*}$,	$r \to 2M^+$	$R^{out}_{\omega l}(r) \sim$	$(r-2M)^{i4M\omega} e^{-i\omega r_*}$,	$r \to 2M^+$
	$e^{-i\omega r_*}$,	$r \to 2M^-$		$e^{4M\omega}(2M-r)^{i4M\omega} e^{-i\omega r_*}$,	$r \to 2M^-$

The quantities $A_{in/out}$ and $B_{in/out}$ in Tables 1 and 2 are complex-valued coefficients. It is worth noticing from Table 2 that the In and Out modes are continuously defined across $r = 2M$.

As it has been mentioned, neither the Schwarzschild nor the IEF coordinate system covers the Maximal Extension of Schwarzschild spacetime. Indeed, in this work we only consider regions I and II of Fig. 1, which are the only regions that are expected to exist in the case of a spherical black hole created from gravitational collapse. We note, however, that it is possible to extend the modes from region I into region I' by mirror-reflecting them with respect to the origin of the Penrose-Carter diagram (Fig. 1). See [16] for more details.

3. Analytical Results

In order to study the separability of Eq. (3), we use the ansatz $\varphi(v, r, \theta, \phi) = e^{-i\omega v}\Psi_{\omega l}(r)\Theta_l(\theta)\Phi_m(\phi)/r$. It thus follows that $\Psi_{\omega l}(r) = e^{-i\omega r_*}R_{\omega l}(r)$. We then obtain from Eq. (3) the following separable partial differential equation:

$$\left(1 - \frac{2M}{r}\right)\frac{\Psi''_{\omega l}}{\Psi_{\omega l}} + \left(\frac{2M}{r^2} - 2i\omega\right)\frac{\Psi'_{\omega l}}{\Psi_{\omega l}} - \frac{2M}{r^3} + \frac{\csc^2(\theta)\Phi''_m}{r^2\Phi_m} + \frac{\Theta''_l}{r^2\Theta_l} + \frac{\cot(\theta)\Theta'_l}{r^2\Theta_l} = 0, \tag{5}$$

where a prime on a function indicates derivative with respect to its argument. This yields the following radial equation:

$$\left(1-\frac{2M}{r}\right)\Psi''_{\omega l}+\left(\frac{2M}{r^2}-2i\omega\right)\Psi'_{\omega l}-\left(\frac{l(l+1)}{r^2}+\frac{2M}{r^3}\right)\Psi_{\omega l}=0. \tag{6}$$

When analysing the singular structure of this equation, we find that $r=0$ and $r=2M$ are regular singular points and $r=\infty$ is an irregular singular point. This means that Eq. (6) has the same singular structure as the radial differential equation in Schwarzschild coordinates.

The boundary conditions of Tables 1 and 2 are given at the singular points $r=2M$ and $r=\infty$. However, we wish to numerically integrate the radial equation by imposing boundary conditions on the solution away from a singular point. For this purpose we use the following local expansions of the radial solutions.

About $r=2M$, we carry out a Fröbenius expansion, thus obtaining two series of the form

$$\Psi_{\omega l}(r)=\sum_{n=0}^{\infty}a_{n\omega l}(r-2M)^{n+\alpha}, \tag{7}$$

where $\alpha=0$ or $4iM\omega$, while the coefficients $a_{n\omega l}$ can be derived by applying the ansatz (7) to Eq. (6). It is clear from Table 1 that $\alpha=0$ corresponds to In modes and $\alpha=4iM\omega$ to Out modes.

Regarding the Up mode, its boundary conditions are set at future null infinity ($\mathcal{J}^+$). Due to the irregularity of this singular point, we consider an ansatz of the form $\Psi_{\omega l}(r)=e^{i\omega r_*}e^{-\mathcal{V}(r)}/r$ and require that $\mathcal{V}(r)\to 0$ as $r\to\infty$ [18]. It is straight-forward to obtain an expansion of $\mathcal{V}(r)$ about $r=\infty$ up to an appropriate order. We note that, although such expansion is only asymptotic, it is nevertheless useful for setting the boundary conditions away from $r=\infty$.

4. Numerical Results

The results presented in this section will be restricted to a particular mode. This mode is $\{l=2,\omega\approx 0.0070/M\}$. All graphs have the same structure: dotted lines represent the real part of a function, while continuous lines represent the imaginary part.

We use the expansion (7) and the boundary conditions as given by Tables 1 and 2. Again, we give special attention to the prescription given in [17] in order to integrate numerically Eq. (6) across the future event horizon.

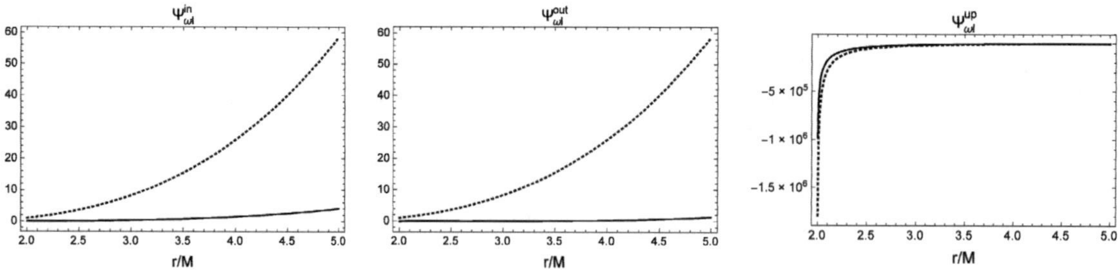

Figure 2. From left to right: In, Out and Up radial functions in the exterior region of the black hole spacetime. Dotted lines represent the real part of a function, while continuous lines represent its imaginary part. We refer the reader to Fig. 3 about the boundedness of the Up radial function.

We also calculated the Wronskian between modes in the exterior region of the black hole. Since only the Wronskian of solutions of the radial equation obtained in Schwarzschild coordinates is constant, we use the relation between $\Psi_{\omega l}(r)$ and $R_{\omega l}(r)$ to obtain the expression

$$W_{Schw}(R^1_{\omega l}, R^2_{\omega l}) = e^{-2i\omega r_*}\left(1 - \frac{2M}{r}\right) W_{IEF}(\Psi^1_{\omega l}, \Psi^2_{\omega l}), \tag{8}$$

where $W(.,.)$ is the Wronskian. From Eq. (8) we have obtained the following values: $W_{Schw}(R^{in}_{\omega l}, R^{up}_{\omega l}) \approx (2.202 + i0.217)10^6 M^{-1}$ and $W_{schw}(R^{out}_{\omega l}, R^{up}_{\omega l}) \approx (2.210 + i0.112)10^6 M^{-1}$.

As previously said, any two types of mode solutions among In, Out and Up form a complete set of mode solutions of Eq.(6). This allows us to express $\Psi^{up}_{\omega l}$ as a linear combination of $\Psi^{in}_{\omega l}$ and $\Psi^{out}_{\omega l}$, with coefficients given by the expression

$$\Psi^{up}_{\omega l} = -\frac{1}{2i\omega} W_{Schw}(R^{out}_{\omega l}, R^{up}_{\omega l})\Psi^{in}_{\omega l} + \frac{1}{2i\omega} W_{Schw}(R^{in}_{\omega l}, R^{up}_{\omega l})\Psi^{out}_{\omega l}. \tag{9}$$

Therefore, using Table 2 and Eq. (7), we are able to obtain In and Out modes in the interior region of the black hole, while Eq. (9) allows us to extend the Up mode across the event horizon. These results can be seen in the graphs of Fig. 3.

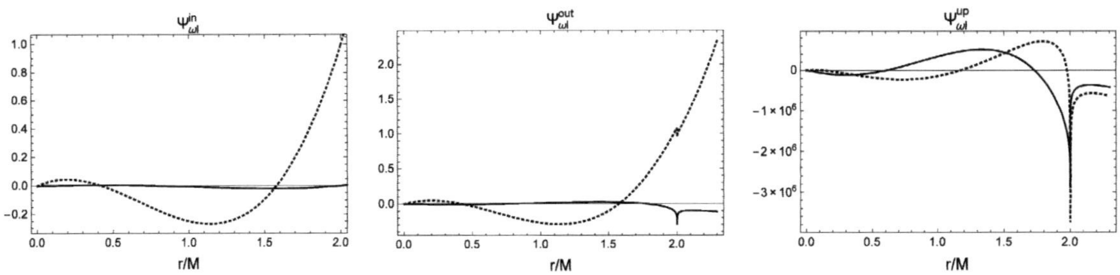

Figure 3. From left to right: In, Out and Up radial functions in the interior region of the black hole. Again, dotted lines represent the real part of a function, while continuous lines represent its imaginary part.

It is interesting to notice that, as expected from the definition of the boundary conditions in Table 2, the In radial function is smoothly defined across the event horizon, while the Out function is continuous, but has a cusp at the event horizon (see Fig. 3). Such feature is carried over to the Up mode via Eq. (9).

We also calculated the Wronskian between the pairs In-Up and Out-Up, this time across the horizon. We found a discontinuity at $r = 2M$, which was expected due to the mentioned cusp in the Out and Up modes. We have obtained the values of $W_{Schw}(R^{in}_{\omega l}, R^{up}_{\omega l}) \approx (2.631 + i0.237)10^6 M^{-1}$ and $W_{Schw}(R^{out}_{\omega l}, R^{up}_{\omega l}) \approx (2.639 + i0.111)10^6 M^{-1}$ inside the horizon.

5. Conclusions

We were able to obtain various mode radial solutions and their Wronskians at high numerical precision in the exterior region of a Schwarzschild black hole. Using a prescription given by Damour and Ruffini [17], we also obtained mode solutions and their Wronskians *inside* the horizon.

Our perspective is to use this work to investigate an expected relationship between quantum modes in the internal and external regions of the Schwarzschild spacetime, as given by the two-point correlation function.

6. Acknowledgments

CB would like to thank Conselho Nacional de Desenvolvimento Científico e Tecnológico for the support given through a MSc scholarship.

Appendix A. Numerical Comparison and Tests

We have thoroughly tested the numerical solutions in order to garantee numerical accuracy. We summarize some of these tests in Table A1.

Table A1. Numerical Tests

Object	Region	Test	Worst Relative Error
$\Psi^{in}_{\omega l}$	Exterior	Series, Equation 7	$\mathcal{O}[10^{-19}]$
$\Psi^{in}_{\omega l}$	Exterior	Numerical Integration similar to [20]	$\mathcal{O}[10^{-29}]$
$\Psi^{in}_{\omega l}$	Exterior	Jaffé Series [18]	$\mathcal{O}[10^{-25}]$
$\Psi^{in}_{\omega l}$	Exterior	Lhs of Equation 6 $= 0$	$\mathcal{O}[10^{-37}]$
$\Psi^{in'}_{\omega l}$	Exterior	Derivative of series, Equation 7	$\mathcal{O}[10^{-17}]$
$\Psi^{up}_{\omega l}$	Exterior	Numerical Integration similar to [20]	$\mathcal{O}[10^{-29}]$
$\Psi^{up}_{\omega l}$	Exterior	Jaffé Series [18]	$\mathcal{O}[10^{-17}]$
$\Psi^{up}_{\omega l}$	Exterior	Lhs of Equation 6 $= 0$	$\mathcal{O}[10^{-37}]$
$W_{Schw}(R^{in}_{\omega l}, R^{up}_{\omega l})$	Exterior	$W'_{Schw}(R^{in}_{\omega l}, R^{up}_{\omega l}) = 0$	$\mathcal{O}[10^{-29}]$
$W_{Schw}(R^{out}_{\omega l}, R^{up}_{\omega l})$	Exterior	$W'_{Schw}(R^{in}_{\omega l}, R^{up}_{\omega l}) = 0$	$\mathcal{O}[10^{-30}]$
$\Psi^{in}_{\omega l}$	Interior	Series, Equation 7	$\mathcal{O}[10^{-21}]$
$\Psi^{in}_{\omega l}$	Interior	Lhs of Equation 6 $= 0$	$\mathcal{O}[10^{-34}]$
$\Psi^{in'}_{\omega l}$	Interior	Derivative of series, Equation 7	$\mathcal{O}[10^{-22}]$
$\Psi^{out}_{\omega l}$	Interior	Numerical Integration similar to [20]	$\mathcal{O}[10^{-19}]$
$\Psi^{out}_{\omega l}$	Interior	Lhs of Equation 6 $= 0$	$\mathcal{O}[10^{-34}]$
$\Psi^{out'}_{\omega l}$	Interior	Derivative of series, Equation 7	$\mathcal{O}[10^{-21}]$
$W_{Schw}(R^{in}_{\omega l}, R^{up}_{\omega l})$	Interior	$W'_{Schw}(R^{in}_{\omega l}, R^{up}_{\omega l}) = 0$	$\mathcal{O}[10^{-29}]$
$W_{Schw}(R^{out}_{\omega l}, R^{up}_{\omega l})$	Interior	$W'_{Schw}(R^{in}_{\omega l}, R^{up}_{\omega l}) = 0$	$\mathcal{O}[10^{-30}]$

References

[1] Teukolsky S A 1973 *Astrophys. J.* **185** 635
[2] Candelas P 1980 *Phys. Rev.* D **21** 2185
[3] Leaver E W 1986 *Phys. Rev.* D **34** 384
[4] Kay B S and Wald R M 1987 *Classical and Quantum Gravity* **4** 893
[5] Whiting B F 1989 *J. Math. Phys.* **30** 1301
[6] Sasaki M and Tagoshi H 2003 *Living Rev. Rel.* **6** 6
[7] Chandrasekhar S and Hartle J B 1982 *Proc. R. Soc. Lond.* A **384** 301
[8] Ori A 1992 *Phys. Rev. Lett.* **14** 2117
[9] Poisson E and Israel W 1990 *Phys. Rev.* D **41** 1796
[10] Schutzhold R and Unruh W G 2010 *Phys. Rev.* D **81** 124033
[11] Parentani R 2010 *Phys. Rev.* D **82** 025008
[12] Hawking S W 1974 *Nature* **248** (5443) 30
[13] Kruskal M D 1960 *Phys. Rev.* **119** 1743
[14] Misner C W *et al* 1973 *Gravitation* (San Francisco: W. H. Freeman and Company)
[15] Frolov V P and Zelnikov A 2011 *Introduction to black hole physics* (Oxford: Oxford University Press)
[16] Frolov V P and Novikov I D 1997 *Black hole physics: basic concepts and new developments*, Fundamental Theories of Physics (Dordrecht: Kluwer Academic Publishers)
[17] Damour T and Ruffini R 1976 *Phys. Rev.* D **14** 332
[18] Casals M and Ottewill A 2013 *Phys. Rev.* D **87** 064010
[19] Philipp D and Perlick V 2015 *Preprint* gr-qc/1503.08101v1
[20] Wardell B *et al* 2014 *Phys. Rev.* D **89** 084021

Polarization of Cosmic Microwave Background

A Buzzelli[1,2], P Cabella[1,2], G de Gasperis[1,2] and N Vittorio[1,2]

[1] Dipartimento di Fisica, Università di Roma "Tor Vergata", via della Ricerca Scientifica 1, I-00133, Roma, Italy

[2] Sezione INFN Roma 2, via della Ricerca Scientifica 1, I-00133, Roma, Italy

E-mail: alessandro.buzzelli@roma2.infn.it

Abstract. In this work we present an extension of the ROMA map-making code for data analysis of Cosmic Microwave Background polarization, with particular attention given to the inflationary polarization B-modes. The new algorithm takes into account a possible cross-correlated noise component among the different detectors of a CMB experiment. We tested the code on the observational data of the BOOMERanG (2003) experiment and we show that we are provided with a better estimate of the power spectra, in particular the error bars of the BB spectrum are smaller up to 20% for low multipoles. We point out the general validity of the new method. A possible future application is the LSPE balloon experiment, devoted to the observation of polarization at large angular scales.

1. Introduction

The Cosmic Microwave Background (CMB) is the relic radiation of the early universe. It is characterized by an almost perfect black body spectrum, with average temperature of about 2.7 K. It is extremely isotropic with temperature fluctuations of only few parts out of 10^5. Nonetheless, the temperature anisotropy pattern has been a fundamental tool to extract cosmological information, since it succeeded in constraining cosmological parameters with very high accuracy. However, to break the degeneracy among some parameters we need to study the (linear) polarization pattern of the CMB, due to Thomson scattering of quadrupolar anisotropies at last scattering surface. Since the polarization is characterized by an amplitude at least one order of magnitude lower than temperature anisotropy, its observation has been an arduous experimental challenge.

Polarization can be decomposed into a gradient component (E-modes) and a curl component (B-modes). While E-modes have been widely observed and analyzed, we still do not have a unique evidence of primordial B-modes. The search for polarization B-modes is perhaps the most exciting and promising field in modern cosmology. We know that scalar (density) perturbations produce only E-modes, while tensor perturbations, i.e. Gravitational Waves (GW), produce both E- and B-modes. Therefore, a detection of B-modes would provide a definitive confirmation to the existence of a stochastic GW background, as predicted by Inflation theory.

Numerous experiments have been done to observe the microwave sky, from ground, balloons and space, each one characterized by its own scanning strategy and experimental setup. In this work we focus on two balloon missions: BOOMERanG (2003) and LSPE, which will be likely launched in 2016.

2. CMB polarization: theory

In this section we briefly review the theoretical and mathematical aspects of CMB polarization. We follow mainly the formalism of [1]. A complete treatment of polarization can be achieved by the two Stokes parameters Q and U, since I describes temperature only and V vanishes, because Thomson scattering induces no circular polarization. We define the polarization tensor P_{ab} as:

$$P_{ab}(\hat{n}) = \frac{1}{2}\begin{pmatrix} Q(\hat{n}) & -U(\hat{n})sin\theta \\ -U(\hat{n})sin\theta & -Q(\hat{n})sin^2\theta \end{pmatrix}.$$

where $\hat{n} = (\theta, \phi)$ is the direction in the sky. Note that this tensor is symmetric, $P_{ab} = P_{ba}$, and trace-free, $g^{ab}P_{ab}$, with the metric given by $g_{ab} = diag(1, sin^2\theta)$.
A generalization of the Helmotz theorem states that any symmetric trace-free rank-2 tensor can be decomposed into a gradient and a curl component. In our case, we refer to these components as E-modes and B-modes, respectively. More explicitly, the polarization tensor can be expanded in terms of basis functions that are gradients and curls of tensor spherical harmonics:

$$\frac{P_{ab}(\hat{n})}{T_0} = \sum_{l=2}^{\infty}\sum_{m=-l}^{l} [a^E_{(lm)}Y^E_{(lm)ab}(\hat{n}) + a^B_{(lm)}Y^B_{(lm)ab}(\hat{n})], \tag{1}$$

where T_0 is the average temperature. The expansion coefficients and the spherical harmonics are given by:

$$a^E_{(lm)} = \frac{1}{T_0}\int d\hat{n} P_{ab}(\hat{n}) Y^{E\,ab*}_{(lm)}(\hat{n}), \tag{2}$$

$$a^B_{(lm)} = \frac{1}{T_0}\int d\hat{n} P_{ab}(\hat{n}) Y^{B\,ab*}_{(lm)}(\hat{n}), \tag{3}$$

$$Y^E_{(lm)ab} = N_l(Y_{(lm):ab} - \frac{1}{2}g_{ab}Y^c_{(lm):c}), \tag{4}$$

$$Y^B_{(lm)ab} = \frac{N_l}{2}(Y_{(lm):ac}\epsilon^c_b + Y_{(lm):bc}\epsilon^c_a), \tag{5}$$

where ':' denotes the covariant derivative, $\epsilon_{ab} = antidiag(\sqrt{g}, \sqrt{-g})$ with g being the determinant of g_{ab}, and the normalization coefficient N_l is:

$$N_l = \sqrt{\frac{2(l-2)!}{(l+2)!}}. \tag{6}$$

From the expansion coefficients we introduce the power spectra as:

$$< a^{X*}_{(lm)} a^{X'}_{(l'm')} >= C_l^{XX'}\delta_{ll'}\delta_{mm'}, \tag{7}$$

where $X, X' = \{E, B\}$. According to the standard cosmological model, $C_l^{TB} = C_l^{EB} = 0$. Recent estimates of C_l^{TT}, C_l^{TE} and C_l^{EE} can be found in [2]. The new frontier of CMB observation is the search for the (primordial) C_l^{BB}.
We already said that B-modes are univocally produced by tensor perturbations. However, we need to consider two important effects that may introduce a B-component in the polarization: gravitational lensing and foreground contamination. The gravitational lensing is caused by large scale mass inhomogeneities between us and the last scattering surface. Its effect is dominant at large multipoles. Foreground contamination is mainly due to synchrotron radiation, below 60 GHz, and dust emission, above 90 GHz. Therefore, the detection of B-modes needs a careful characterization of these effects, in order to isolate the B-component of inflationary origin.
The exact level of B-mode signal is still uncertain and it depends on the Inflation potential. Qualitatively, we have $C_l^{TT} \simeq 10C_l^{TE} \simeq 10^2C_l^{EE} \simeq 10^4C_l^{BB}$.

3. CMB polarization: data analysis

The extraction of cosmological information from observations is a complex procedure and it depends on the type of the experiment (ground based, balloons, satellites). However, the general pipeline is the same for any experimental setup and it can be separated into three main steps. An experiment observes the sky and collects data in a given temporal sequence: we refer to these data as TOD (Time Ordered Data). The TOD must be calibrated, i.e. expressed in μK, and the pointing must be reconstructed. The first step of the pipeline is the projection of the TOD on the sky, that means to build a sky map (map-making). This implies a compression from $10^{10} - 10^{9}$ TOD to $10^{7} - 10^{6}$ pixels of a map. The map-making is the most time consuming part of the data analysis. Therefore, an important task is to find a computationally efficient way to perform the analysis that preserves all the relevant cosmological information. A detailed description of map-making can be found in [3] and [4]. The second step of the pipeline is the estimation of the angular power spectra from the maps, that implies a compression to $10^{4} - 10^{3}$ data. In general, the sky maps contain "bad" pixels, i.e. pixel not observed or contaminated. For example, balloon experiments are designed to observe only a fraction of the sky, and even all-sky missions, like Planck [5], have to remove part of the sky from the analysis, due to foreground contamination. Therefore, we have first to estimate the spectra referred to the maps (pseudo-spectra) and then statistically extend the spectra to the celestial sphere (full sky-spectra) through appropriate algorithms. To do that, in our analysis we used the HEALPix software [6] and the MASTER algorithm [7], respectively. The third step is the computation of the $\simeq 10$ cosmological parameters, such as the expansion rate and the curvature of the universe, and the relative abundances of baryons, Dark Matter and Dark Energy, from the spectra. This can be done with the COSMOMC software [8].

Our study is focused mainly on the first two steps. We know that the detectors of a CMB experiment may be affected by strong cross-correlations, i.e. the noise of each detector is the sum of a non-correlated component and a common-mode component seen by all the detectors [9]. Our aim is to test the improvement that we gain in the estimate of the spectra when we take into account the cross-correlated noise component, usually neglected in past works, in the map-making code.

The TOD from one detector can be placed in a vector d, defined as:

$$d = Pm + n. \tag{8}$$

where the operator P is the "pointing matrix", containing all the information about the scanning strategy, the vector m is the map and the vector n is the instrumental noise. In general, m refers to the Stokes parameters I, Q and U. Some assumptions are made for the noise [9], in particular we assume to know all its statistical properties and we take it as stationary. In case of more than one detector, Eq. (8) is still valid, with the quantities from each detector being concatenated end to end.

Map-making means to solve Eq. (8) for m. A good estimator $\tilde{m}$ of m can be found by a Generalized Least Squares approach:

$$\tilde{m} = (P^T N^{-1} P)^{-1} P^T N^{-1} d. \tag{9}$$

where N is the "noise covariance matrix" defined as $N =< nn^T >$. Eq. (9) can be iteratively solved by the method of conjugate gradients.

To include cross-correlation, we write the noise of the i-th detector at time t as [9]:

$$n_t^i = \tilde{n}_t^i + \gamma^i c_t, \tag{10}$$

where $\tilde{n}_t^i$ is the uncorrelated term, c_t is the correlation, dependent only on time, and γ^i is an amplitude, dependent only on the detector. Generally, for each detector, the spectrum of the noise has the following shape:

$$P(f) = A\left[1 + \left(\frac{f_k}{f}\right)^{\alpha}\right], \tag{11}$$

where f is the frequency, and A, f_k (knee frequency) and α are constants. We have then a $1/f$ part at low f and a constant part at high f.

4. BOOMERanG (2003)

BOOMERanG is a telescope mounted on a stratospheric long duration balloon [10–12], whose main goal is the accurate measurement of the sky in the microwave band. Two missions were launched: in 1998 and in 2003, both from Antarctica. We focus on the second mission, aimed at the polarization detection. The observation strategy was designed to cover a small fraction of the sky with "deep" integration to observe polarization, a wider fraction of the sky with "shallow" integration to observe temperature, and a portion of the Galactic plane (Fig. 1).

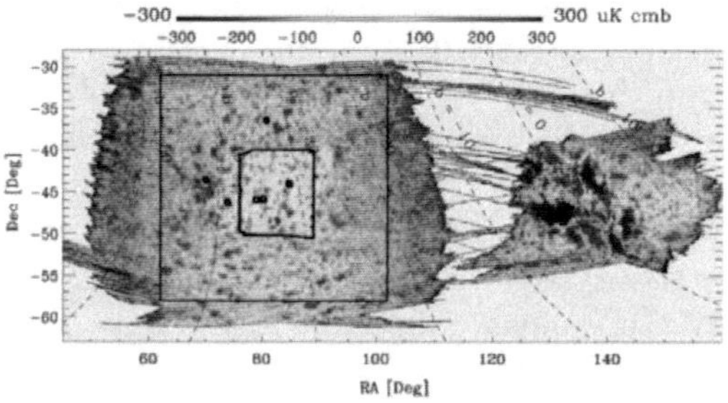

Figure 1. Regions scanned by BOOMERanG (2003): the deep region inside the small square, the shallow region inside the big square, and the portion of Galactic plane on the right [13].

All the survey was carried out in three frequency bands centered at 145, 245 and 345 GHz. For the 145 GHz band, there are four pairs of detectors; both the other bands have four single detectors (Fig. 2). We concentrate our analysis on the 145 GHz band.

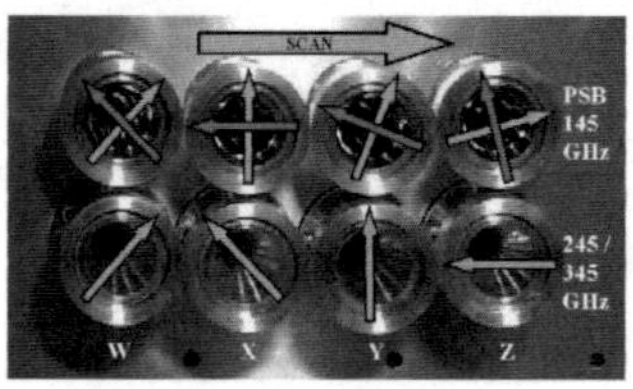

Figure 2. Focal plane of BOOMERanG (2003): four pairs of detectors (Polarization Sensitive Bolometers) at 145 GHz and four single detectors at 245 and 345 GHz [14].

We applied the procedure described in the previous section to 200 Monte Carlo simulations of BOOMERanG (2003) data. In particular we used the map-making code ROMA [15], extended with a possible cross-correlation among the detectors. To include the effect of the cross-correlated noise component, we added two contributions to the detector auto-noise: the cross-correlated noise between detectors in the same pair and the one among each of the other six detectors. These contributions share with the auto-noise the shape $(1/f)^2$ at low f, but they are characterized

by a constant level at high f of 10% and 1% of the auto-noise, respectively.
Once generated the maps, we estimated the power spectra. In Fig. 3 we compare the BB power spectrum obtained neglecting and including cross-correlation, using data from the deep region.

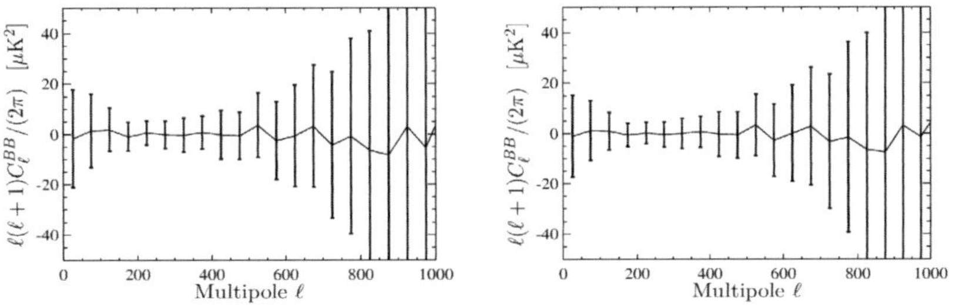

Figure 3. Comparison between the BB power spectra estimated neglecting (on the left) and including (on the right) the cross-correlated noise component.

To estimate the effect of considering the cross-correlated noise component, we analysed the ratio of the spectra error bars estimated with and without cross-correlation. We expect that neglecting the cross-correlation will bring to overestimate the noise contribution for low multipoles.

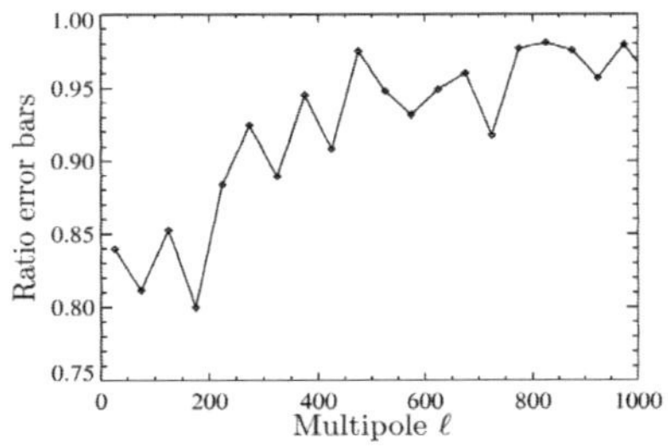

Figure 4. Ratio between the BB spectra error bars, when including or not the noise cross-correlation.

From Fig. 4 we notice that the new algorithm has provided smaller error bars on the BB spectrum. In particular, the errors are reduced by 20% for low multipoles. Since the BB spectrum is still buried into instrumental noise, the reduction of the error bars represents a very important step towards a first detection of polarization B-modes.

5. LSPE

The techniques shown in last section have general validity. We applied the map-making code to simulated data of BOOMERanG (2003) and estimated the power spectra, showing the improvement obtained by taking into account the cross-correlation among the detectors. The same pipeline can be applied to other microwave experiments, in particular a suitable future application is the LSPE (Large Scale Polarization Explorer) balloon mission [16]. Its launch is expected for the end of 2016. It is aimed at observing CMB polarization at large angular scales with a sensitivity about ten times greater than Planck's. The primary target is to constrain the B component of polarization whose information is encoded in the low multipoles. We said that

B-modes represent a unique signature of tensor perturbations produced during Inflation. For this purpose, LSPE is expected to improve the limit on the ratio of tensor to scalar perturbations amplitudes down to $r = 0.03$. A detection of r would constrain the energy scale of Inflation according to the relation $E = 3.3 \times 10^{16} r^{1/4} GeV$. Moreover, LSPE will produce wide maps of foreground polarization generated in our Galaxy by synchrotron and interstellar dust emission. During its circumpolar Arctic flight, LSPE will observe a large fraction of the northern sky (25% of the whole sky) with angular resolution of 1.5 degrees FWHM. The payload will host two instruments: an array of bolometric polarimeters, SWIPE [17], which will map the sky at 95, 145 and 245 GHz, and an array of coherent polarimeters, STRIP [18], which will survey the same sky region at 43 and 90 GHz. The multi-frequency information provided by LSPE would allow to characterize the foreground contamination, in order to disentangle the single polarized components. Our next step will be the generation of simulated maps in order to estimate the possible improvement that the new algorithm will provide for LSPE. Our interest will be focused on the SWIPE instrument, characterized by two focal planes with a total number of 110 detectors for frequency. We expect that the problem of cross-correlated noise will be a crucial issue in the data analysis of LSPE.

6. Conclusions

In this work we tested a new code for data analysis of CMB polarization, which takes into account a possible cross-correlated noise component among the detectors. We applied this algorithm to simulated data of BOOMERanG (2003) and we showed the benefit obtained in the estimate of the power spectra. In particular, we reduced the BB spectrum error bars up to 20% at low multipoles. The whole pipeline tested on BOOMERanG can be entirely applied to the data analysis of the forthcoming LSPE mission. Its goal will be mainly the characterization of polarization at large angular scales, where the information of B-modes is encoded. The reduction of the error bars achieved with the new code could be crucial in the search for polarization B-modes with LSPE.

Acknowledgments

We thank P. de Bernardis, S. Masi and the BOOMERanG collaboration for providing us with the observational dataset of the BOOMERanG (2003) experiment and K. M. Gorski et al. for the use of the HEALPix package.

References

[1] Cabella P and Kamionkowski M 2005 arXiv:astro-ph/0403392v2
[2] Ade P A R *et al* [Planck collaboration] arXiv:1502.01589 [astro-ph.CO]
[3] Natoli P *et al* 2001 *A&A* **372** 346-356
[4] Natoli P *et al* 2009 *Mem. S. A. It. Suppl. Vol.* **13** 84
[5] Adam R *et al* [Planck collaboration] arXiv:1502.01582v2 [astro-ph.CO]
[6] Gorski K M *et al* 2005 *Astrophys. J.* **622** 759
[7] Hivon E *et al* 2002 *Astrophys. J.* **567** 2
[8] Lewis A and Bridle S 2002 *Phys. Rev. D* **66** 103511
[9] Patanchon G *et al* 2008 *Astrophys. J.* **681** 708-725
[10] Masi S *et al* 2006 *A&A* **458** 687-716
[11] MacTavish C J *et al* 2006 *Astrophys. J.* **647** 799-812
[12] Montroy T *et al* 2003 *New Astron. Rev.* **47** 1057-1065
[13] Piacentini F *et al* 2007 *New Astron. Rev.* **51** 244-249
[14] http://oberon.roma1.infn.it/boomerang/b2k/focal_plane.html
[15] de Gasperis G *et al* 2005 *A&A* **436** 1159-1165
[16] Aiola S *et al* 2012 *SPIE Conf. Series* **8446** 277 arXiv:astro-ph/12080281v1
[17] de Bernardis P *et al* 2012 *SPIE Conf. Series* **8452** 125 arXiv:astro-ph/12080282v1
[18] Bersanelli M *et al* 2012 *SPIE Conf. Series* **8446** 279 arXiv:astro-ph/12080164v1

Journal of Physics: Conference Series **689** (2016) 012004 doi:10.1088/1742-6596/689/1/012004

Reconstruction of the local inflationary potential with different correlation levels

A Di Marco[1], P Cabella[1], N Vittorio[1]

[1] Physics Department, Tor Vergata University - Cosmology Group - Via della Ricerca Scientifica, 1, 00133 Italy

E-mail: alessandro.di.marco@roma2.infn.it

Abstract. We review the puzzles of the standard Big Bang model and cosmic inflation as their possible solutions. The relation between inflation and the spectra of the cosmological perturbations is emphasized. In particular we focus on the local reconstruction of the shape of the inflationary potential from observations and the consequences of a direct detection of cosmological gravitational waves, exploring different correlation levels between the spectral index n_s and the tensor-to-scalar ratio r of the primordial perturbations.

1. Introduction

The standard Big Bang model explains important features of the universe: the Hubble expansion, the abundances of light elements (H, He, ...) and the presence and thermal nature of the Cosmic Microwave Background (CMB). Nevertheless, there are some observational evidences that remain unexplained inside this picture, the so-called puzzles of the standard cosmology [1–5]. They can be synthesized in the following questions:

- The flatness problem: why the universe looking back in time is so flat?
- The horizon problem: why the universe shows the same statistical properties also in non causally connected regions?
- The monopoles problem: where are the magnetic monopoles that were produced during the breaking of the Grand Unified Theory?

In addition, there is the problem of the origin of cosmic structures: in the standard scenario there is not a natural way to introduce the perturbations that lead to the formation of the large scale structures as well as the anisotropies in the CMB. The simplest solution to resolve these problems is to introduce an accelerated expansion of the early universe about 10^{-35} s after the initial singularity, on energy scales of the order of 10^{15} GeV [1–3], often called inflation. Mathematically, inflation is defined by $\ddot{a}(t) > 0$, where $a(t)$ is the scale factor of the universe, and it acts by producing:

- An early stretching of the constant-time hypersurfaces inducing the flatness.
- A causal horizon greater then the horizon in the standard hot big bang context.
- A dilution of the magnetic monopoles on the largest scales, reducing the possibility of their detection.

2. Inflationary cosmology and scalar fields

The simplest method to introduce an accelerated phase is to consider a neutral and homogeneous scalar field called inflaton ϕ that dominates the early times and that is minimally coupled to gravity with canonical kinetic term [1–3]. The inflaton field is also characterized by some potential $V(\phi)$ with an initial plateau or quasi plateau region, as shown in Figure 1. The cosmological action for early times becomes, in natural units $c = \hbar = 1$:

$$S = \int d^4x\sqrt{-g}\left\{\frac{1}{16\pi G}R - \frac{1}{2}g_{\mu\nu}\partial^{\mu}\phi\partial^{\nu}\phi - V(\phi)\right\} \tag{1}$$

where R is the Ricci scalar, $g_{\mu\nu}$ is the metric tensor, g the determinant of the metric tensor and G the Newton constant. The Friedmann equations derived from this action are therefore:

$$\left(\frac{\dot{a}}{a}\right)^2 = \frac{8\pi G}{3}\rho, \qquad \frac{\ddot{a}}{a} = -\frac{8\pi G}{3}(\rho + 3p) \quad \text{and} \quad H(t) = \frac{\dot{a}}{a} \tag{2}$$

where H is the Hubble rate while the energy density and the pressure of the scalar field are:

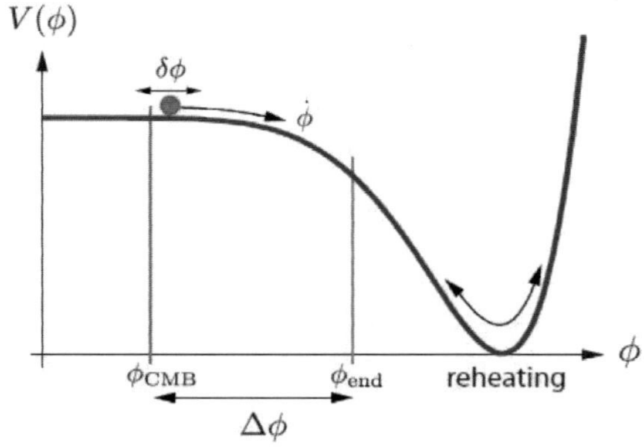

Figure 1. In the simplest scenario a scalar field is the energy source for the inflationary process.

$$\rho(\phi) = \frac{1}{2}\dot{\phi}^2 + V(\phi), \quad p(\phi) = \frac{1}{2}\dot{\phi}^2 - V(\phi). \tag{3}$$

What happens is shown in Fig.1. When inflation starts the scalar field moves slowly through the quasi plateau region of the potential. Here the kinetic term is subdominant with respect to the potential one, i.e $\dot{\phi}^2 << V(\phi)$, thus $\rho + 3p < 0$, realizing the acceleration $\ddot{a}(t) > 0$ in the second Friedmann equation. Later, the inflaton relaxes toward the global minimum, the expansion reduces and the final stage called reheating starts. The field oscillates and decays producing entropy and normal particles [6]. The entire mechanism can be codified by the so called slow roll parameters defined as:

$$\epsilon = -\frac{\dot{H}(t)}{H^2(t)}, \qquad \eta = -\frac{\ddot{H}(t)}{H(t)\dot{\phi}(t)} \quad \text{or} \quad \epsilon_V = \frac{1}{2}M_p^2\left(\frac{V'(\phi)}{V(\phi)}\right)^2, \qquad \eta_V = M_p^2\left(\frac{V''(\phi)}{V(\phi)}\right). \tag{4}$$

In ϵ_V and η_V the $'$ represents the derivative with respect to the scalar field and in both cases M_p is the reduced Planck mass. In the slow roll limit $\dot{\phi}^2 << V(\phi)$, one has:

$$\epsilon \simeq \epsilon_V, \qquad \eta = \eta_V - \epsilon_V. \tag{5}$$

Inflation takes place when $\epsilon < 1$ or $\epsilon_V \ll 1$.

3. Inflation and quantum fluctuations

In inflationary cosmology the problem of the origin of structures is also resolved [2,3]. In the early universe the scalar field is dominant and from a quantum mechanics point of view it is characterized by quantum fluctuations $\delta\phi(x,t)$ with zero mean value in a macro time scale, by definition:

$$\phi(t) \rightarrow \phi(x) = \overline{\phi}(t) + \delta\phi(x,t).$$

Such fluctuations imply fluctuations on the stress-energy tensor $T_{\mu\nu}$ and by the Einstein equations, on the metric tensor $g_{\mu\nu}$. The accelerated expansion stretches these fluctuations on astronomical scales and here they freeze-out with a non-zero mean value in a macro time scale and become classical perturbations. These perturbations will induce the anisotropies of the CMB and will grow into the large scale structure of the universe. There are two useful quantities to describe a given cosmological perturbation field f:

- The power spectrum, which describes the presence of the perturbation on a given scale. It is defined as: $P(k) = \frac{k^3}{2\pi^2}|f(k)|^2$, where k is the wave number.
- The spectral index, which describes the variation of the perturbation with respect to the scale. It is defined as: $n(k) = \frac{dP(k)}{d\ln k}$.

Inflation excites scalar, vector and tensor modes [2,3]. The power spectrum and the index of scalar and tensor sectors at first order in slow roll expansion are [5–7]:

$$P_s(k) = \frac{1}{8\pi^2 M_p^2}\frac{H^2}{\epsilon}, \quad n_s = 1 - 4\epsilon + 2\eta \quad \text{and} \quad P_t(k) = \frac{2}{\pi^2}\frac{H^2}{M_p^2}, \quad n_t = -2\epsilon \tag{6}$$

and it is possible to define the so called tensor-to-scalar ratio of the perturbation amplitudes:

$$r = 16\epsilon = -8n_t. \tag{7}$$

The vector sector is uninteresting because the related modes decay as the universe expands [2].

4. Reconstruction of the inflationary potential

The common way to study the inflationary phase consists in writing a lagrangian density with a given self-interaction potential, extracting the values for (n_s, r) using Eq. (4) and (5), and comparing them to cosmological data. This is particularly simple in the case of a canonical single field model. An alternative solution uses directly the data sets to constrain, at least locally, the shape of the inflationary potential [8]. This approach is justified by two pieces of evidence:

- The observable modes are stretched on astronomical scales (over the Hubble radius $R_H = c/H$, with c speed of light) about 60 e-folds before the end of inflation, when the inflaton field is on the plateau.
- The value of the inflaton field during this phase is approximately constant, ϕ_0.

In this sense it is possible to perform a Taylor series expansion of the potential, around ϕ_0 and connect the coefficients of the expression with the inflationary observables:

$$V(\phi) = V(\phi_0) + V'(\phi_0)(\phi - \phi_0) + \frac{1}{2}V''(\phi_0)(\phi - \phi_0)^2 + \dots \quad .$$

To do this one needs an appropriate equation for the potential. Such expression is the so called Hamilton-Jacobi equation for the inflationary dynamics in which the independent variable is the inflaton field instead of cosmic time [7,8]:

$$V(\phi) = 3M_p^2 H^2(\phi) - 2M_p^2 H'^2(\phi). \tag{8}$$

In the Hamilton formalism it is possible to define a complete hierarchy of Hubble slow roll parameters (HSRP) [7,8]:

$$\epsilon(\phi) = 2M_p^2\left(\frac{H'(\phi)}{H(\phi)}\right)^2, \quad \beta_n(\phi) = 2M_p^2\left(\frac{(H')^{n-1}H^{(n+1)}}{H^n}\right)^{1/n} \quad n \geq 1 \tag{9}$$

where the symbol $(n+1)$ indicates the derivative of order $n+1$ with respect to the field. These quantities codify order after order the dynamics of inflation. Using Eq. (8) and Eq. (9), and passing to the observables P_s, n_s and r by Eq. (6), at first order approximation in HSRP and up to quadratic order in $\Delta\phi$, one has[1]:

$$V(\phi) = \frac{3}{2}\pi^2 M_p^4 P_s(k) r|_{\phi_0} + \frac{3}{4\sqrt{2}}\pi^2 M_p^3 P_s(k) r^{\frac{3}{2}}|_{\phi_0}\Delta\phi + \frac{\pi^2 M_p^2 P_s(k) r}{2}\left[\left(\frac{9r}{16}\right) - \frac{3}{2}(1-n_s)\right]|_{\phi_0}\Delta\phi^2. \tag{10}$$

The derivation of this expression is given in [8].

5. Sampling the potential: results and conclusions

In this work we present the general results for the reconstruction of the potential both at first and second order in slow roll expansion, considering:

- Direct detection of inflationary gravitational waves, i.e. primordial tensor perturbations.
- Different possible correlation levels for the couple (n_s, r).

In particular we assume that future observations will provide a Gaussian distribution both for n_s and r with a given correlation ρ. The assumed central value and standard deviation of the two distributions are $n_s = 0.968$, $\sigma_{n_s} = 0.006$ and $r = 0.05$, $\sigma_r = 0.01$ while $P_s = 20.08 \times 10^{-10}$ for $k = 0.05$ Mpc^{-1}, compatible with the current Planck 2015 data [9]. In the first order framework we can normalize the expansion respect the 0-th order term and define:

$$V(\phi) = V(\phi_0)\left[1 + d_1\Delta\phi + \frac{1}{2}d_2\Delta\phi^2\right], \quad d_1 = \frac{1}{2}\sqrt{\frac{r}{2}} \quad \text{and} \quad d_2 = \frac{1}{3}\left[\frac{9r}{16} - \frac{3}{2}(1-n_s)\right].$$

Sampling N times the coefficients d_1 and d_2 using the previous statistics, and assuming $\rho = 0.1$ and $\rho = 0.5$ we get the results in Figures 2 and Figure 3. These plots show that in presence of higher correlation levels the distributions tend to compactify. This effect is reasonable because a more accentuate correlation constrains the ranges that the single variables can assume individually. For completeness we also report the results using the next order approximation in the slow roll parameters using expressions derived in [8]. In this case the square of the third-slow roll parameter $\xi^2(\phi)$ plays an important role and hence its physical counterpart, the running of the scalar spectral index defined as $\alpha_s = \frac{dn_s}{d\ln k}$. We assume a Gaussian distribution for it with central and standard deviation values given by $\alpha_s = -0.0084$, $\sigma_\alpha = 0.0082$, compatible with Planck data [9]. The results are showed in Figure 4 and Figure 5. In this case $d1$ and $d2$ are not normalized to the 0-th order term of the Taylor series expansion. The shapes of the distributions are quite different from the previous ones because of the different and more complex expressions of d_1 and d_2. However, they show the same trend for different values of ρ. We can conclude this preliminary work summarizing some conclusions:

- The technique of reconstruction is an important tool to understand which order tends to dominate in the Taylor expansion.
- The technique is particularly sensitive to the available statistics: in the future a detection of r can improve the results (for an example, see [10]).
- The next step is to reproduce the previous analysis using an higher order Taylor expansion in the inflationary variables to get more information about the physics around ϕ_0.

[1] In this work we assume that $\dot{\phi} < 0$ so $\sqrt{\epsilon} = +\sqrt{2}M_p H'/H$ and the first order term is positive.

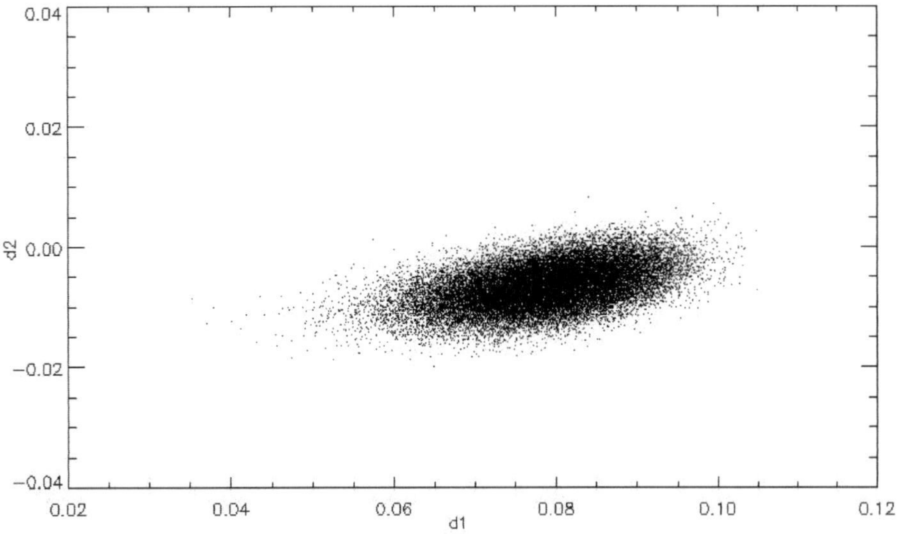

Figure 2. Distribution of d2 vs d1 at 1st order with $\rho = 0.1$ and N=30000.

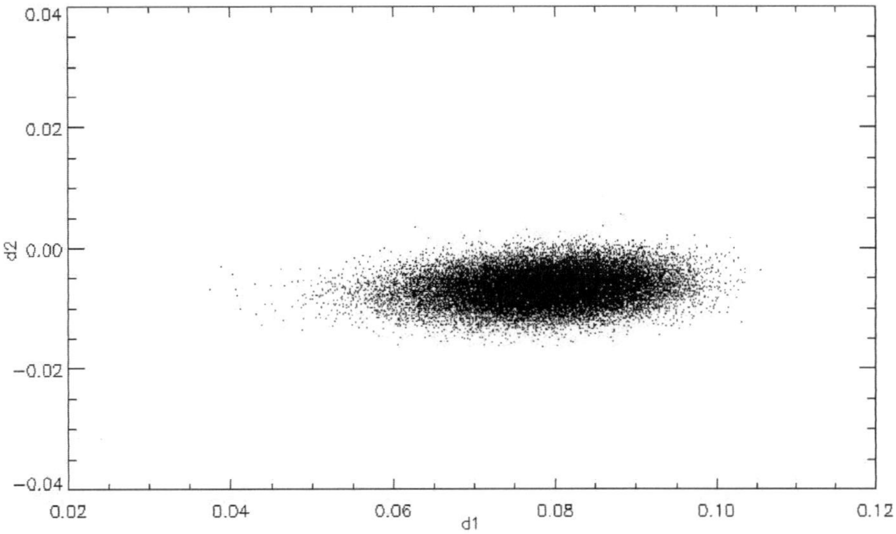

Figure 3. Distribution of d2 vs d1 at 1st order with $\rho = 0.5$: in this case the cloud of samples is denser with respect to the previous case.

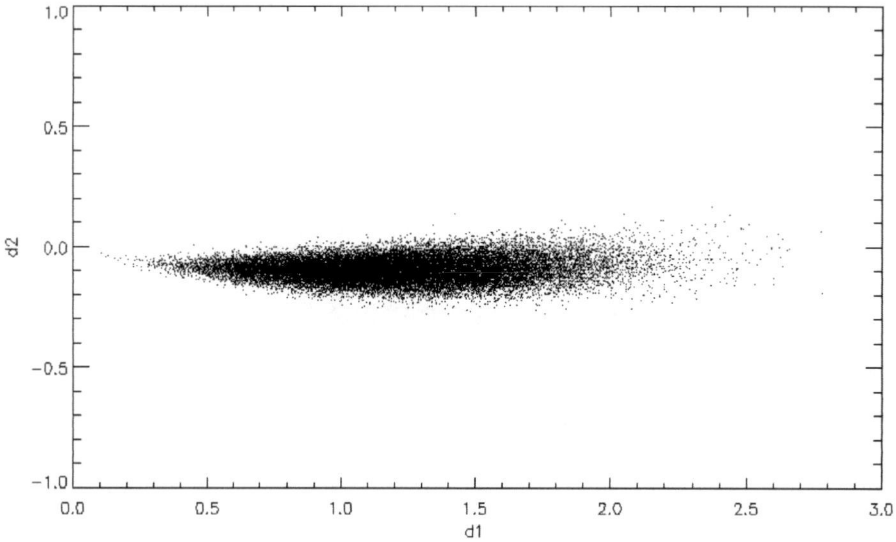

Figure 4. Distribution of d2 vs d1 at 2nd order with $\rho = 0.1$.

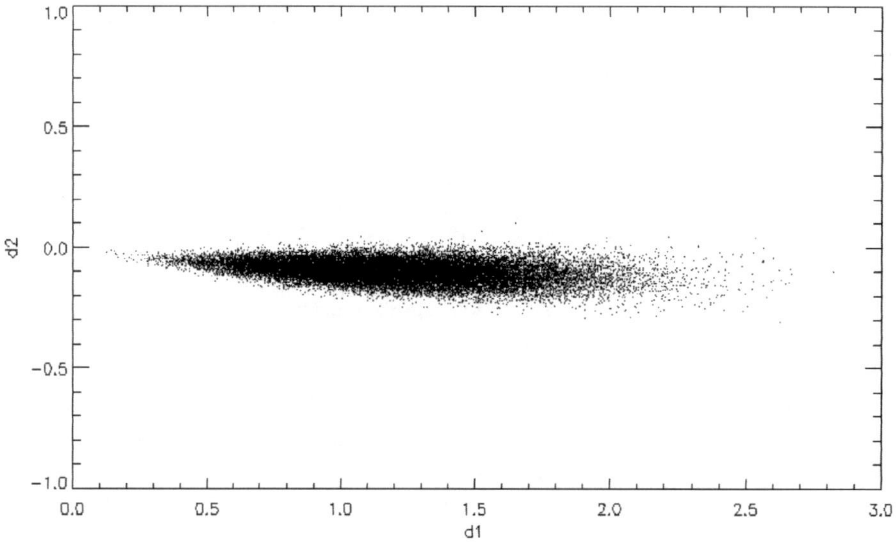

Figure 5. Distribution of d2 vs d1 at 2nd order with $\rho = 0.5$: the distribution becomes thinner.

References

[1] Guth A 1981 *Phys. Rev.* D **23** 2
[2] Weinberg S 2008 Cosmology (Oxford: Oxford University Press)
[3] Dodelson S 2003 Modern Cosmology (San Diego: Accademic Press)
[4] Linde A D 1981 *Phys. Lett.* B **108** 6 389-393
[5] Linde A D 1983 *Phys. Lett.* B **129** 3 4
[6] Albrecht A, Steinhardt P J, Turner M S, Wilczek F 1982 *Phys. Rev. Lett.* **48** 20
[7] Liddle A R, Parsons P, and Barrow J D 1994 *Phys. Rev.* D **50** 7222
[8] Lidsey J E, Liddle A R, Kolb E W, Copeland E J, Barreiro T and Abney M 1997 *Rev. Mod. Phys.* **69** 373
[9] Planck 2015 results. XIII. Cosmological parameters submitted to 2015 A&A (arXiv:1502.01589)
[10] Yin-Zhe Ma, Wang Y 2014 JCAP (09) 041

Journal of Physics: Conference Series **689** (2016) 012005 doi:10.1088/1742-6596/689/1/012005

A Detailed Study of the Mass Distribution of the Galaxy Cluster RXC J2248.7−4431

Gabriel B. Caminha[1], **Piero Rosati**[1], **Claudio Grillo**[2] **and the CLASH-VLT team**

[1]Dipartimento di Fisica e Scienze della Terra, Università degli Studi di Ferrara, Via Saragat 1, I-44122 Ferrara, Italy

[2]Dark Cosmology Centre, Niels Bohr Institute, University of Copenhagen, Juliane Maries Vej 30, DK-2100 Copenhagen, Denmark

E-mail: gbcaminha@fe.infn.it

Abstract. In this work we use strong gravitational lensing techniques to constrain the total mass distribution of the galaxy cluster RXC J2248.7-4432 (RXC J2248, $z_{lens} = 0.348$), also known as Abell S1063, observed within the Cluster Lensing And Supernova survey with Hubble (CLASH). Thanks to its strong lensing efficiency and exceptional data quality from the VIsible Multi-Object Spectrograph (VIMOS) and Multi Unit Spectroscopic Explorer (MUSE) on the Very Large Telescope, we can build a parametric model for the total mass distribution. Using the positions of the multiple images generated by 7 multiply-lensed background sources with measured spectroscopic redshifs, we find that the best-fit parametrisation for the cluster total mass distribution is composed of an elliptical pseudo-isothermal mass distribution with a significant core for the overall cluster halo, and of truncated pseudo-isothermal mass profiles for the cluster galaxies. This model is capable to predict the positions of the multiple images with an unprecedented precision of $\approx 0''.3$. We also show that varying freely the cosmological parameters of the ΛCDM model, our strong lensing model can constrain the underlying geometry of the universe via the angular diameter distances between the lens and the sources and the observer and the sources.

1. Introduction

The concordance ΛCDM cosmological model predicts that, on large scales, the baryonic matter is $\approx 5\%$ of the present energy density of the Universe, and a poorly understood component called dark matter constitutes $\approx 20\%$. A more exotic component, the dark energy Ω_Λ with negative pressure (i.e., with an equation of state of the form $P = w\rho$, where P and ρ are the pressure and the density, respectively, and w is a negative quantity), is responsible for the current accelerated expansion of the Universe [1–3]. On galaxy clusters scales (≈ 1 Mpc) the dominant component is the dark matter, constituting $\approx 90\%$ of the cluster total mass. The rest is composed by hot gas, galaxies and the intra cluster light.

Gravitational lensing effects, i.e. distortion, magnification and creation of multiple images from the same background source, depend on the mass distribution of the midway object acting as a gravitational lens, and on the angular diameter distances between the observer, the lens, and the background source. Note that this effect does not depend on the dynamical state of the lens, making it a powerful probe of the total mass distribution of the lenses and the underlying geometry of the Universe. On galaxy cluster scales, only recently it has been possible to exploit the observed positions of spectroscopically confirmed families of multiple images to obtain precise measurements of the total mass distributions in the core of these lenses [4,5] and the first constraint on cosmological parameters [6].

In this work we constrain the total mass distribution of the galaxy cluster RXC J2248 ($z_{lens} = 0.348$) using the images of multiply-lensed background sources. We take advantage of the CLASH multiband *HST* data and extensive spectroscopic information that we have collected on the cluster members and background, lensed sources in this galaxy cluster with the VIMOS and MUSE instruments on the VLT [7,8].

2. RXC J2248

RXC J2248 was first identified as Abell S1063 in [9] and its high mass and redshift makes it a powerful gravitational lens. It was one of the 25 clusters observed within CLASH [10] in 16 filters, from the UV through the NIR, with the ACS and WFC3 cameras onboard *HST*. These photometric data allow us to identify strong lensing features, such as giant arcs, multiple image families and distorted background sources.

As part of the CLASH-VLT Large program (ID 186.A−0798, P.I.: P. Rosati), the cluster RXC J2248 was observed with the VIMOS spectrograph between June 2013 and May 2015. The inner regions of the galaxy cluster were observed in a total exposure time of 15 hours with 16 masks, 12 with the low-resolution blue grism (spectral resolution of 28Å and wavelength range of 3700 − 6700Å) and 4 with the intermediate resolution grism (spectral resolution of 13Å and wavelength range 4800 − 10000Å). Observations with the new integral-field spectrograph MUSE on the VLT were conducted in the South-West part of the cluster as part of the MUSE science verification program (ID 60.A-9345, P.I.: K. Caputi). A 8520 second total exposure was obtained in June 2014 with a seeing of $\approx 1''$. The MUSE data cube covers 1×1 arcmin^2, with a pixel size of $0''.2$, over the wavelength range 4750 − 9350Å and a dispersion of 1.25 Å. All the details of the VLT and MUSE observations and data reduction can be found in [7] and [8], respectively. From these spectroscopic data, we identified 7 bona-fide spectroscopically confirmed multiple image families. These families do not suffer of systematic effects that we cannot take into account in our parametric model, such as lensing by line of sight mass structures and strong effects of very close cluster members. A discussion of the impact of these effects is given in [11].

3. Strong lensing modelling

We use the strong lensing observables to reconstruct the total mass distribution of RXC J2248. The positions of the multiple images from a single background source depend on the relative distances (observer, lens and source) and on the total mass distribution of the intervening lens. We describe below our methodology to determine the mass distribution of the cluster from the observed positions of the identified multiple images.

3.1. Mass distribution components

In view of its regular shape, we model the total mass distribution of the lens as the sum of three main components: 1) a smooth component describing the extended dark matter, 2) the brightest cluster galaxy (BCG) and 3) small scale halos associated to galaxy members.

For the smooth mass component (intra cluster light, hot gas and, mainly, dark matter) we adopt a Pseudo Isothermal Elliptical Mass Distribution (hereafter PIEMD) [12]. This model has 6 parameters: the centre position (x_0 and y_0), the ellipticity and its orientation angle (ε and θ), the fiducial velocity dispersion (σ_v) and the core radius (r_{core}). The PIEMD parametrization describes well cluster mass distributions in strong lensing studies, and sometimes provides a better fit than the canonical Navarro-Frenk-White (hereafter NFW) [13,14] mass distribution. In the reference [5], using a similar high-quality data set, it has been found for example that

Journal of Physics: Conference Series **689** (2016) 012005 doi:10.1088/1742-6596/689/1/012005

the dark matter components of the galaxy cluster MACS J0416.1−2403 are better described by PIEMD models.

The galaxy members selection is performed following the method adopted in [5] (see Section 3.3.1 of the referenced paper). Essentially, we investigate the loci of a large sample of spectroscopically confirmed cluster members and field galaxies in a multidimensional color space. We then model the probability density distributions (PDFs) of cluster member and field galaxy colours as multidimensional gaussians, with means and covariances determined using the Minimum Covariance Determinant [15]. We label each galaxy as being a cluster member or a field galaxy, using the determined PDFs in a Bayesian hypotheses inference. We then classify galaxies using a probability threshold that is a good compromise between purity and completeness, and thus selecting additional cluster members with no spectroscopic redshifts. With this procedure we select a total of 139 cluster members in the model within a distance of $1'$ from the BCG, 64 of which are spectroscopically confirmed. Each cluster member is modelled as a truncated PIEMD [16,17] with zero ellipticity and core radius, and a finite truncation radius. Following a standard procedure in cluster-scale strong lensing analyses, we reduce the number of free parameters describing the cluster members assuming a constant total mass-to-light ratio of cluster members. Thus, we are left with only two parameters, the reference velocity dispersion σ_v^{gals} and truncation radius r_{trunc}^{gals}, describing the overall properties of all cluster members.

Due to a generally different formation history, the BCG is often observed to significantly deviate from this scaling relation [18]. We therefore introduce two additional free parameters associated to the BCG (σ_v^{BCG} and r_{trunc}^{BCG}), keeping its position fixed at the center of the light distribution. We eventually have a total of 10 free parameters to describe the total mass distribution of RXC J2248.

3.2. Lensing modelling definitions

The strong lensing modelling is performed using the public software *lenstool* [19, 20]. Once the model components are defined, the best-fit model parameters are found by minimizing the distance between the observed and predicted positions of the multiple images, and the parameter covariance is quantified using a Bayesian Markov Chain Monte Carlo (MCMC) technique. In detail, to find the best-fitting model, we define the lens plane χ^2 function as follows:

$$\chi^2(\vec{\Pi}) := \sum_{j=1}^{N_{\rm fam}} \sum_{i=1}^{N_{\rm im}^j} \left(\frac{\left| \vec{\theta}_{i,j}^{\rm obs} - \vec{\theta}_{i,j}^{\rm pred}\left(\vec{\Pi}\right) \right|}{\sigma_{i,j}^{\rm obs}} \right)^2 , \tag{1}$$

where $N_{\rm fam}$ and $N_{\rm im}^j$ are the number of families and the number of multiple images belonging to the family j, respectively. $\vec{\theta}^{\rm obs}$ and $\vec{\theta}^{\rm pred}$ are the observed and the predicted positions of the multiple images, and $\sigma^{\rm obs}$ is the uncertainty in the observed position. The predicted position of an image is a function of both lens parameters and cosmological parameters, all represented by the vector $\vec{\Pi}$. In this work we adopt flat priors on all parameters. The set of parameters $\vec{\Pi}$ that provides the minimum value of the χ^2 function (χ^2_{min}) is called the best-fitting model, while the predicted positions of this model are referred to as $\vec{\theta}^{\rm bf}$.

3.3. The family ratio

The availability of a large number of multiple images with measured spectroscopic redshifts and a relatively regular mass distribution, makes RXC J2248 a suitable cluster lens to test the ability to constrain cosmological parameters. Strong lensing is sensitive to the underlying geometry of the Universe via the angular diameter distances between the observer and the lens (D_{OL}), the observer and the source (D_{OS}), and the lens and the source (D_{LS}). For one source, the lens equation can be written as

$$\vec{\theta} = \vec{\beta} + \frac{D_{LS}}{D_{OS}} \hat{\vec{\alpha}} \left(\vec{\theta} \right), \tag{2}$$

where $\vec{\theta}$ and $\vec{\beta}$ are the angular positions on the lens and source planes, respectively, $\hat{\vec{\alpha}}$ is the deflection angle and the cosmological dependence is embedded into the angular diameter distances. The ratio between the cosmological distances is degenerated with the parameters of the lens mass distribution. However, when a significant number of multiply lensed sources at different redshifts is present, this degeneracy can be broken and a leverage on cosmological parameters can be obtained via the so called family ratios:

$$f_k(\vec{\pi}) = \frac{D(\vec{\pi})_{LS,1} D(\vec{\pi})_{OS,2}}{D(\vec{\pi})_{LS,2} D(\vec{\pi})_{OS,1}}, \tag{3}$$

where $\vec{\pi}$ is the set of cosmological parameters and 1 and 2 are two different sources at redshifts z_{s1} and z_{s2}.

This technique has been successfully applied to the galaxy clusters Abell 2218 [21] and Abell 1689 [6]. In this work, we use the ΛCDM cosmological model, which includes as free parameters the energy density of the total matter of the universe (ordinary and dark matter) Ω_m, the dark energy density Ω_Λ and the equation of state parameter of this last component, $w = P/\rho$. All results from our lens model are summarized in the next Sections.

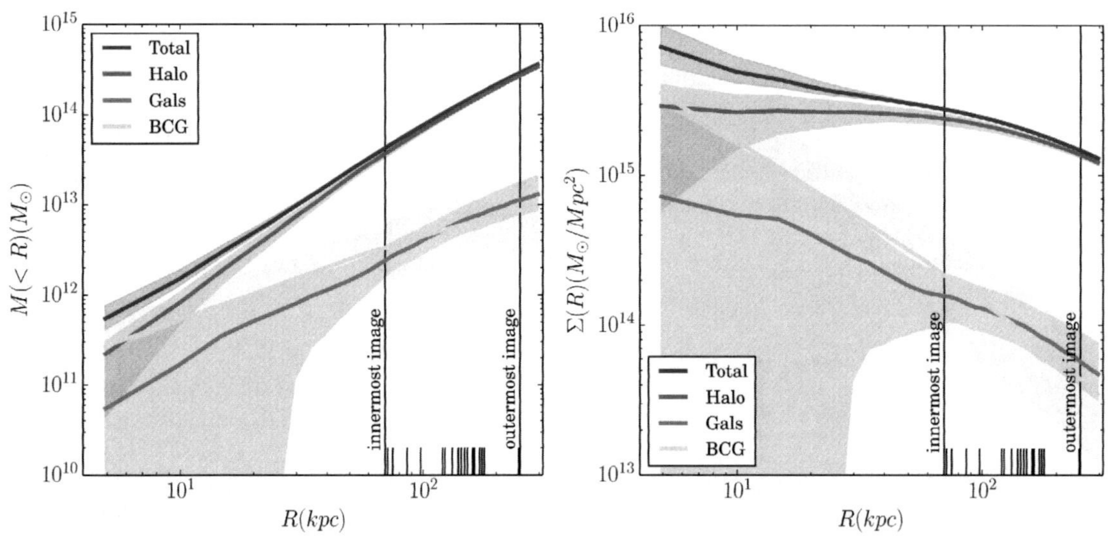

Figure 1. Projected mass, $M(<R)$, and surface mass density $\Sigma(R)$ relatively to the BCG center for the total, cluster dark-matter (Halo), galaxy members (Gals) and BCG mass components. The solid lines represent the median and the filled contours the 68% confidence level from the strong lensing total mass reconstruction. The vertical lines show the positions of the multiple images.

4. Results

We first consider a model fixing the cosmological parameters to $\Omega_m = 0.3$, $\Omega_\Lambda = 0.7$ and $w = -1$, i.e., a flat cosmology with a cosmological constant. We use this model to study the overall properties of the cluster total mass distribution. Secondly, we let the cosmological parameter free, together with those parametrizing the total mass distribution. In this case, we consider a flat ΛCDM model (i.e. $\Omega_\Lambda = 1 - \Omega_m$), and varying Ω_m and w. The parameters of this model are constrained via the family ratios (see Equation 3). In both cases, the mean offset between the observed and predicted positions of the images is $\approx 0''.3$, showing that our models can reproduce the observations with high accuracy.

4.1. Constraining the mass distribution

In the left panel of Figure 1, we show the reconstructed projected mass as a function of the radius, considering the BCG centre as reference, using our strong lensing model with fixed cosmological parameters. The solid lines are the medians and the filled regions are the 68% confidence levels for different mass components: total, dark-matter (Halo), galaxy members (Gals) and BCG. The vertical black lines indicate the distance of the multiple images to the BCG center. The larger statistical uncertainties on the recovered mass components, especially for the BCG and the galaxy members, in the innermost regions of the cluster (i.e. $R < 30\,\mathrm{kpc}$) are a consequence of the lack of multiple images in these regions. However, these uncertainties decrease significantly in the outskirts, $R > 50\,\mathrm{kpc}$, as expected due to the presence of the multiple image constraints. We find that, the total projected mass within 250 kpc is $2.90^{+0.03}_{-0.03} \times 10^{14} M_\odot$ (the errors are given by the 68% confidence level) in our model, higher than the values of $2.68^{+0.03}_{-0.05} \times 10^{14} M_\odot$ and $2.67^{+0.08}_{-0.08} \times 10^{14} M_\odot$ presented in [22] and [23], respectively. Although these measurements are not consistent within the statistical errors, the mean values do not differ by more than 10%, and are likely due to different assumptions in these studies. We show the surface mass density given by our model in the right panel of Figure 1. It is interesting to note that the dark matter halo component, green line and filled regions, shows a relatively flat density up to $R \approx 100\,\mathrm{kpc}$.

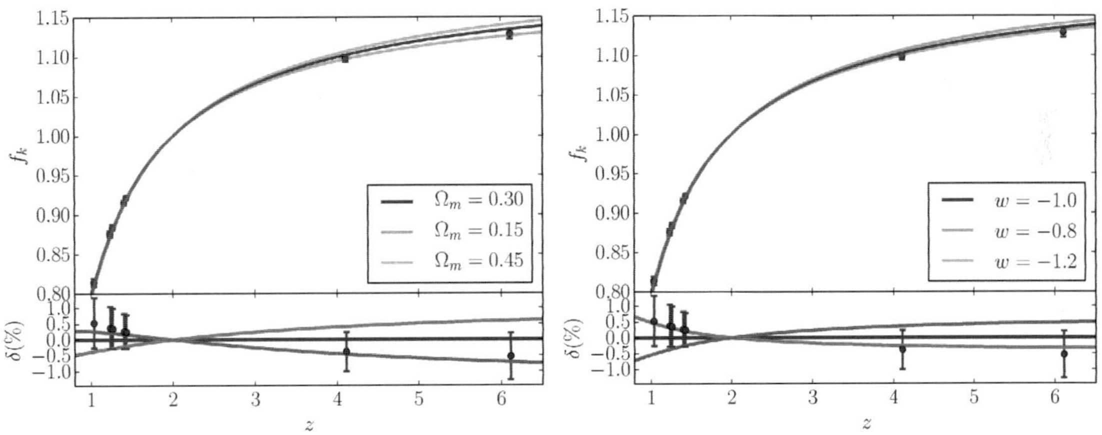

Figure 2. Top panels: family ratios f_k of our strong lensing modelling with free cosmology as a function of redshift, renormalized for $z_{s2} = 2$. The blue circles show the best fit values from the lens model and the error bars are the 68% confidence level. The continuous lines show the the family ratios with fixed values of the cosmological parameters. Bottom panels: relative difference with respect to the concordance flat ΛCDM model (i.e. $\Omega_m = 0.3$ and $w = -1$).

4.2. Recovering the family ratios f_k

In Figure 2, we show the family ratios of the observed sources renormalized for a source at $z_{s2} = 2$ (see Equation 3), from our best-fitting strong lensing model with free cosmological parameters. In the top panels, the blue circles and the error bars are the median and the 68% confidence level and the continuous lines the prediction of the ΛCDM model with different values of Ω_m (left panel) and w (right panel). In the bottom panels we show the relative difference δ with respect to the concordance flat ΛCDM model (i.e. $\Omega_m = 0.3$ and $w = -1$). The statistical scatter given by the error bars is a result of degeneracies between f_k and the mass distribution parameters. Our modelling seems to favour models with lower values of w and Ω_m when comparing with the concordance model, however the error bars do not exclude this cosmological model. Moreover, the wide range of the source redshifts, from 1.0 to 6.1 is critical to yield good constraints on the evolution of f_k up to high values of z. A complete study of the cosmological constraints using the galaxy cluster RXC J2248 is presented in [11].

5. Conclusions

Using the positions of spectroscopically confirmed multiply lensed images, we obtain a detailed reconstruction of the total mass distribution of the galaxy cluster RXC J2248. Although the scatter is larger for the mass components of the galaxy members and the BCG in the innermost regions, $R < 30$kpc, in the outer regions these components are well constrained. We show that RXC J2248 is particularly suitable for constraining the background geometry of the Universe with strong lensing modelling due to its unique combination of a regular shape and a large number of multiple images spanning a wide redshift range, as revealed by our spectroscopic survey.

References

[1] Riess A G *et al.* 1998 *AJ* **116** 1009
[2] Eisenstein D J *et al.* 2005 *ApJ* **633** 560
[3] Planck Collaboration 2014 *A&A* **571** A16
[4] Halkola A *et al.* 2008 *A&A* **481** 65–77
[5] Grillo C *et al.* 2015 *ApJ* **800** 38
[6] Jullo E *et al.* 2010 *Science* **329** 924
[7] Balestra I *et al.* 2013 *A&A* **559** L9
[8] Karman W *et al.* 2015 *A&A* **574** A11
[9] Abell G O *et al.* 1989 *ApJS* **70** 1
[10] Postman M *et al.* 2012 *ApJS* **199** 25
[11] Caminha G B *et al.* 2015 *ArXiv:1512.04555*
[12] Kassiola A and Kovner I 1993 *ApJ* **417** 450
[13] Navarro J F *et al.* 1996 *ApJ* **462** 563
[14] Navarro J F 1997 *ApJ* **490** 493–508
[15] Rousseeuw P J 1984 *JASA* **79** 871
[16] Elíasdóttir A *et al.* 2007 *ArXiv:0710.5636*
[17] Suyu S H and Halkola A 2010 *A&A* **524** A94
[18] Postman M *et al.* 2012 *ApJ* **756** 159
[19] Kneib J P *et al.* 1996 *ApJ* **471** 643
[20] Jullo E *et al.* 2007 *New Journal of Physics* **9** 447
[21] Soucail G *et al.* 2004 *A&A* **417** L33
[22] Johnson T L *et al.* 2014 *ApJ* **797** 48
[23] Monna A *et al.* 2014 *MNRAS* **438** 1417

A new approach to the variability characterization of active galactic nuclei

R Middei[1], F Vagnetti[1], M Antonucci[1], R Serafinelli[1]

[1] Dipartimento di Fisica, Università di Roma "Tor Vergata", via della Ricerca Scientifica 1, 00133 Roma, Italy

E-mail: riccardo.middei@gmail.com

Abstract. The normalized excess variance is a popular method used by many authors to estimate the variability of active galactic nuclei (AGNs), especially in the X-ray band. We show that this estimator is affected by the cosmological time dilation, so that it should be appropriately corrected when applied to AGN samples distributed in wide redshift intervals. We propose a formula to modify this estimator, based on the use of the structure function. To verify the presence of the cosmological effect and the reliability of the proposed correction, we use data extracted from the XMM-Newton Serendipitous Source Catalogue, data release 5 (XMMSSC-DR5), and cross-matched with the Sloan Digital Sky Survey quasar catalogue, of data release 7 and 12.

1. Introduction

Active galactic nuclei are bright sources shining both in the local and the high redshift Universe. These objects share a number of characteristic properties among which variability. Flux variations occur in the whole electromagnetic spectrum and act on many timescales, in the case of the X-ray domain, from fraction of a day up to several years. Studies concerning the AGN variability in the X-ray domain are of great importance in order to constrain the physics of the inner part of these sources. Often variability studies have been produced using the so-called normalised excess variance (NXS) estimator (e.g. [1, 2]), while only few ensemble studies based on the structure function (SF) are available, e.g. [3]. The NXS is a very popular method but it has been pointed out that it depends on the duration of the light curves [4]. We demonstrate that this method can not be used in a naive way. It is affected by the cosmological time dilation, so that, when used for a wide range of redshifts, it can lead to erroneous variability estimations.

2. Variability estimators

The NXS provides a straightforward way to characterize the mean variability of a single source. It is defined by the equation:

$$\sigma_{NXS}^2 = \frac{S^2 - \sigma_n^2}{\langle f \rangle^2} , \quad (1)$$

where $\langle f \rangle = \sum_{i=1}^{N} f_i/N$ is the mean flux computed over the available flux measures f_i of the same source, $S^2 = \sum_{i=1}^{N}(f_i^2 - \langle f \rangle^2)/N$ is the total variance of the light curve, while σ_n^2 is the mean square photometric error associated to the measured fluxes f_i. The σ_{NXS}^2 is the mean quadratic variation corrected for the error, so that it estimates the intrinsic variability. It can be used for approximate variability characterizations even for sources with only two measurements of the flux at different epochs. When more data are available its estimate becomes more precise (see Fig. 2).

On the other hand, the structure function analysis can be used to describe the variability of a single source as a function of the time lag between observations, but it requires a large number of data, of the order of some hundreds, to properly address the variability at the different time lags. The power of this tool is instead unveiled when the SF is used to provide ensemble variability characterizations. In both cases, the structure function describes variability as a function of the time lag τ between two observations as expressed in the formula:

$$SF(\tau) = \sqrt{\langle [\log f_X(t+\tau) - \log f_X(t)]^2 \rangle - \sigma_n^2} , \quad (2)$$

where σ_n^2 is the mean quadratic error as in Eq. 1, and the mean can be computed on a sample of sources for ensemble analyses, or on the different flux measurements of the same source for individual studies. It is found that the SF increases with the time lag without evidence of any flattening up to few years, and the increase can be approximated by a power law function [3]:

$$SF \propto \tau^{\beta} , \quad (3)$$

where $\beta = 0.10 \pm 0.01$ has been estimated.

As introduced in the previous section, the NXS estimator depends on the duration of the light curve. This implies that, if we compare variability of several sources using NXS we have to choose a fixed time span for all the light curves. When variability of sources in a wide redshift interval is investigated, light curves can still have different durations in their rest-frames. This is due to the well known effect of the cosmology, already pointed out in [5]. It leads to the relation $\Delta t_{rest} = \Delta t_{obs}/(1+z)$. This formula shows that even if in our reference frame we select only equally long light curves, they will not have the same lengths in the rest frame of the sources. Fig. 1 explains how the cosmological time dilation acts on the light curves, indeed, it shows how their length changes as a function of the redshift. Considering that sources have different redshift values, they all present light curves with different length in their own rest-frame.

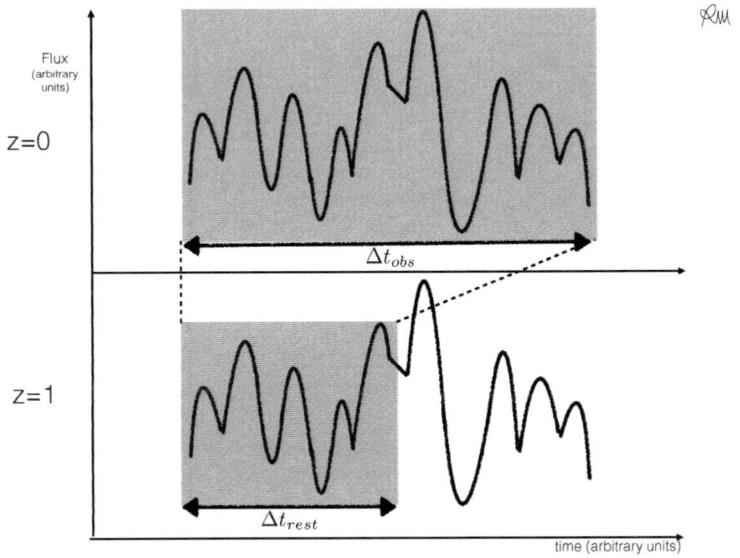

Figure 1. The effect of redshift on NXS variability estimates is shown. In the top panel it is drawn a light curve that belongs to a quasar located at $z = 0$ and, for comparison in the bottom panel, that of a quasar in a farther region of the Universe ($z = 1$). The intrinsic light curve is the same in both cases but, as emphasized by the shadowed area, its observed part changes. In the case of the local source it is detected a larger amplitude in flux variations that provides a greater σ_{NXS} value. Conversely, when the object is located at high redshift, it is observed a smaller amplitude in flux variations, with a smaller value of NXS.

Considering the dependence on the lag between the observations described by the SF, we expect that the NXS itself will depend on the duration of the monitoring time in the rest-frame (which for our data spans between days and years, see Sect. 3) through the integrated effect of the SF within such interval. In fact, neglecting the photometric noise both in Eq. 1 and 2, we find

$$\sigma^2_{NXS} = S^2 / \langle f \rangle^2 \approx \langle (\delta \log f_X)^2 \rangle / (\log e)^2 = \langle SF^2 \rangle / [2(\log e)^2] \,, \qquad (4)$$

where the factor 1/2 accounts for two independent measurements contributing to each flux difference, and the average is computed over the rest-frame interval corresponding to the observed monitoring time. Using Eq. 3, finally we find:

$$\sigma^2_{NXS} \propto \Delta t^{2\beta}_{rest} \,. \qquad (5)$$

3. Data Sample

Quasar variability studies are based on the comparison between at least two flux measurements of the same source, thus a suitable data set has to contain multiple

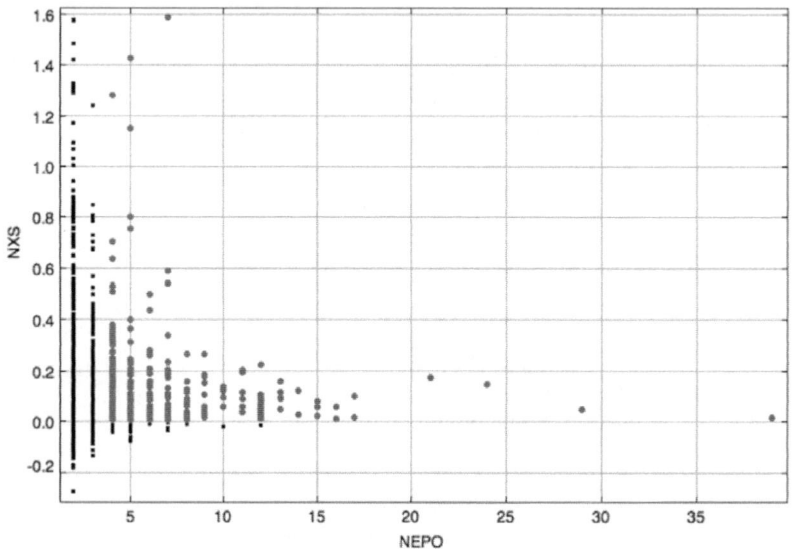

Figure 2. NXS estimates of our sources as a function of the number of epochs (NEPO). Points represent the NXS values computed. Notice that when the NXS parameter is computed using a larger number of measurements its values are distributed in a narrower region. Black points refer to NXS estimates computed with less than four observations and/or to negative values for NXS. Red points constitute the subsample we use in the analysis and are characterized by positive NXS values obtained with at least four observations.

observations for several sources. The XMM-Newton Serendipitous Source Catalogue (XMMSSC DR5) [6] contains a large number of X-ray detections (565,962) for 396,910 single sources achieved between 2000 and 2013. In this catalogue to each observation is associated a warning flag (SUM_FLAG), summarizing its photometric quality. We discard all detections labelled with SUM_FLAG>2 i.e. suggesting that the observation has a high probability to be artifact and unreliable. To extract only quasar observations we take advantage of the 7th [7] and the 12th [8] data releases of the Sloan Digital Sky Survey (SDSS) quasar catalogues. These catalogues as well as the XMMSSC DR5 contain for each observation the equatorial coordinates, α and δ. This allows us to cross-match the catalogues by coordinates, thus to extract only quasar data from the serendipitous X-ray detections. We consider a match when an XMMSSC observation lies within 5 arcsec in both α and δ from an observation listed in any of the two SDSS quasar catalogues.

The joint data set contains only spectroscopically confirmed quasars, and as a further step we require that sources have at least two observations. This leads us to obtain a preliminary sample of 2,700 sources (all the points in Fig. 2). Referring to Eq. 1, it might be possible to obtain $\sigma^2_{NXS} < 0$ when the mean squared error is larger than the variance. In such cases we discard the NXS estimates, because the variability information provided is embedded into the noise, so reducing the sample to 1,775 sources. Furthermore, since the NXS parameter is an averaged estimate of the source variability, it is desirable to increase the number of measurements used to compute NXS. The computed σ^2_{NXS} will better approximate the intrinsic NXS of the source when a large number of data is used. However, only few sources are observed several times so that, in this work we used a subsample containing 416 sources observed at least four times, the red points in Fig. 2.

4. Trend and correction

We report in Fig. 3 (blue points) the NXS estimates computed according to Eq. 1. The linear least squares fit to these estimates shows how the cosmic time dilation acts on the NXS, yielding a $\sigma^2_{NXS} \propto \Delta t_{rest}^{0.21\pm0.04}$. The Pearson correlation coefficient has a value $r = 0.24$, while the probability to find by chance the same correlation is $P(> r) = 6 \times 10^{-7}$. It is possible to correct the NXS estimates using the ensemble

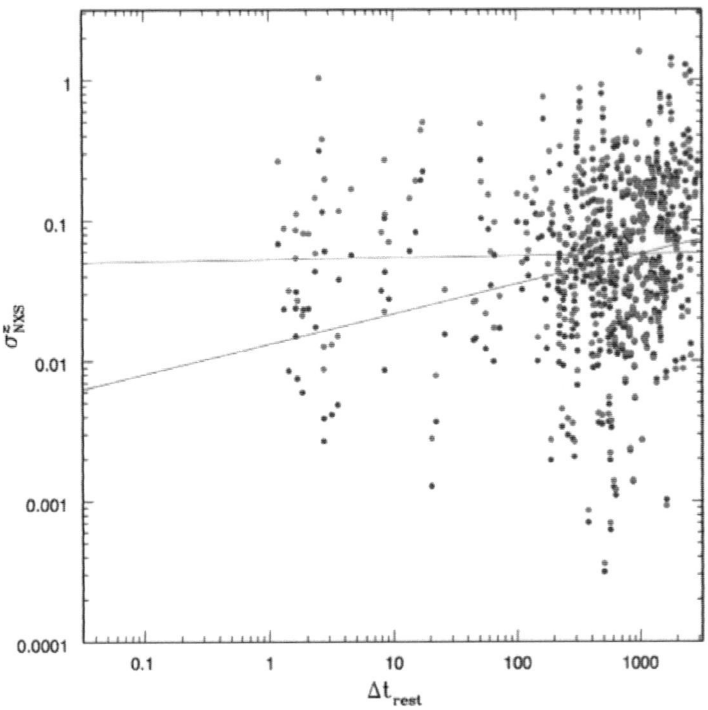

Figure 3. Comparison of the corrected (red) and uncorrected (blue) NXS estimates and their least squares fits as functions of the rest-frame light curve duration Δt_{rest}.

description provided by the structure function. The SF ensemble analysis, indeed, characterizes the variability of a large number of sources as a function of the lag τ between two observations. Eq. 5 describes how σ^2_{NXS} depends on the duration of the light curve in the rest-frame, Δt_{rest}, and it allows us to extrapolate the expected variability for a certain source at a certain time lag. In order to compare the variability of sources characterized by shorter light curves with those objects having longer light curves, we normalize Eq. 5 to a given time interval Δt^* that can be freely chosen:

$$\sigma^{2\,*}_{NXS} = \sigma^2_{NXS} \left(\frac{\Delta t^*}{\Delta t_{rest}}\right)^{2\beta} = \sigma^2_{NXS} \left(\frac{\Delta t^*}{\Delta t_{obs}}\right)^{2\beta} (1+z)^{2\beta} \,. \quad (6)$$

We adopt $\Delta t^* = 1000$ days, and the results are shown in Fig. 3 (red points).

Applying the correction described in Eq. 6, the formal fit to the $\sigma^{2\,*}_{NXS}$ values has slope $a \sim 0.001$ and Pearson correlation coefficient $r = 0.06$, with $P(> r) = 0.83$, indicating that the two quantities are not anymore correlated.

The correction prevents the NXS wrong estimates for high redshift sources, whose shorter Δt_{rest} can be extrapolated to the same Δt^* as for the other sources actually measured at such long times.

5. Final Remarks

The NXS method provides a reliable variability description only when applied to local sources, otherwise it leads to variability underestimates. This is due to the cosmological time dilation that shrinks the duration of the light curves. The correction we propose takes advantage of the statistical characterization provided by ensemble SF-based studies. With this correction it is possible to extrapolate the NXS estimates to a fixed rest-frame duration of the light curve Δt^*, so obtaining corrected values which can be compared to each other.

Acknowledgements

We thank the organisers of the sixth Young Researchers Meeting. We thank the Gran Sasso Science Institute for hospitality. This research has made use of data obtained from the 3XMM XMM-Newton serendipitous source catalogue compiled by the 10 institutes of the XMM-Newton Survey Science Centre selected by ESA.

References

[1] Nandra K, George IM, Mushotzky FR, et al. 1997, *ApJ*, **476**, 70
[2] Ponti G, Papadakis I, Bianchi S, et al. 2012, *A&A*, **542**, A83
[3] Vagnetti F, Turriziani S and Trevese D, 2011, *A&A*, **536**, A84
[4] Lawrence A, and Papadakis I, 1993, *ApJ*, **414**, L85
[5] Gaskell CM, 1981, *Astrophys. Lett.*, **21**, 103
[6] Rosen SR, Webb NA, Watson MG, et al. 2015, `ArXiv: 1504.07051`
[7] Schneider DP, Richards GT, Hall PB, et al. 2010, *Astronomical Journal*, **139**, 2360
[8] Paris I, Petitjean P, Aubourg E, et al. 2014, *A&A*, **563**, A54

Journal of Physics: Conference Series **689** (2016) 012007 doi:10.1088/1742-6596/689/1/012007

Ensemble spectral variability study of Active Galactic Nuclei from the XMM-Newton serendipitous source catalogue

R Serafinelli[1], F Vagnetti[1], R Middei[1]

[1]Dipartimento di Fisica, Università di Roma "Tor Vergata", via della Ricerca Scientifca 1, 00133 Roma, Italy

E-mail: roberto.serafinelli@roma2.infn.it

Abstract. The variability of the X-Ray spectra of active galactic nuclei (AGN) usually includes a change of the spectral slope. This has been investigated for a small sample of local AGNs by Sobolewska and Papadakis [1], who found that slope variations are well correlated with flux variations, and that the spectra are typically steeper in the bright phase (*softer when brighter* behaviour). Not much information is available for the spectral variability of high-luminosity AGNs and quasars. In order to investigate this phenomenon, we use data from the XMM-Newton Serendipitous Source Catalogue, Data Release 5, which contains X-Ray observations for a large number of active galactic nuclei in a wide luminosity and redshift range, for several different epochs. This allows to perform an ensemble analysis of the spectral variability for a large sample of quasars. We quantify the spectral variability through the *spectral variability parameter* β, defined by Trevese and Vagnetti [2] as the ratio between the change in spectral slope and the corresponding logarithmic flux variation. We find that the spectral variability of quasars has a *softer when brighter* behaviour, similarly to local AGNs.

1. Introduction

Active galactic nuclei are very bright extragalactic objects, located in the dynamical center of their host galaxy, powered by the accretion of matter around a supermassive ($M_{BH} = 10^6 \div 10^9 M_{\odot}$) black hole, along a geometrically thin, optically thick disk. This disk is believed to be responsible of the optical/UV emission, while a hot ($T \simeq 10^8 \div 10^9$ K) corona is believed to produce hard X-ray photons by comptonization of the disk photons. This emission is responsible for the observed power-law like spectrum in the X-ray band.

Variability can be a powerful instrument to probe for this physical effect, because it can provide information about both shape and physical behaviour of the corona, which are far from being completely understood.

In addition to the study of amplitude variability, it is also interesting to investigate the spectral variability of active galactic nuclei. This feature has already been studied in the optical/UV band by many authors, (e.g. [2], [3], [4]) and a *harder when brighter* behaviour of a power-law spectrum $F \propto \nu^{\alpha}$ has been observed. Trevese and Vagnetti [2] defined a *spectral variability parameter* β:

$$\beta = \frac{\Delta\alpha}{\Delta \log F}\,. \qquad (1)$$

relating variations in flux and spectral index α.
In the X-ray band the opposite behaviour (*softer when brighter*) has been found for some individual sources (e.g. [5], [6] and more recently [7]). One of the few systematic studies of X-ray spectral variability has been performed by Sobolewska and Papadakis [1], who found such behaviour for 10 nearby Seyfert galaxies, hinting that most of the sources can match the scenario of the superposition of a constant reflection component that overlaps a power-law component, variable in both amplitude and slope.
Currently there have been no systematic studies concerning the spectral variability of quasars and high-luminosity type-1 AGNs, one notable exception being [8], in which the authors find evidence of spectral variability for a fraction of the sources, also with a softer when brighter trend.

2. The Data Set

We are looking for the *softer when brighter* effect also in type-1 luminous AGNs and quasars. In order to select this kind of sources, we have cross-matched the XMM-Newton Serendipitous Source Catalogue (XMMSSC-DR5) with two quasar catalogues from the Sloan Digital Sky Survey, i.e. SDSS-DR7Q and SDSS-DR12Q. These two optical catalogues are complementary and therefore both must be included, so that the catalogue we are going to analyze could be as much complete as possible. SDSS-DR7Q lists 105, 783 spectroscopically confirmed quasars [9] and SDSS-DR12Q lists 297, 301 quasars [10]. Both catalogs do not include active galactic nuclei classes such as Seyfert galaxies, BL Lacertae and type-2 quasars. Therefore, they are perfect for our goal.
XMMSSC-DR5 contains 565, 962 observations taken between 2000 and 2013 [11], corresponding to 396, 910 unique sources, which means that many sources have been observed more than once. Since we are dealing with variability issues, we required that the sources must be observed multiple times, at least twice. Therefore we only selected sources with multi-epoch observations.

Table 1. Number of observations and sources listed for all the catalogues used in this work

Catalogue	Number of observations	Number of sources
SDSS-DR7Q	/	105, 783
SDSS-DR12Q	/	297, 301
XMMSSC-DR5	565, 962	396, 910
Cross-matched catalogue	7, 837	2, 700

Both SDSS catalogues and XMM-Newton Serendipitous Source Catalogue list equatorial coordinates α and δ for each observation. We included in our catalogue every XMMSSC observation that lies within 5" from any catalogued source in any of the two SDSS quasar catalogues. In the cases in which we found the same source in both SDSS catalogues, we chose the one contained in SDSS-DR12Q, because it is more recent.
Something that we had to take into account is that not all of the XMMSSC photometric observations are optimal: the 'goodness' of the observation is quantified by a parameter of the catalogue, labeled as `SUM_FLAG`. We excluded from our list the worst photometric observations, meaning all observations with `SUM_FLAG` > 2 and we obtained a final catalogue that has 7, 837 X-ray observations, corresponding to 2, 700 quasar sources.
The number of observations and the number of sources listed in each catalogue are summarized in Table 1.

Journal of Physics: Conference Series **689** (2016) 012007 doi:10.1088/1742-6596/689/1/012007

3. β Parameter and Hardness Ratios

We quantify the spectral variability of a source by means of the *spectral variability parameter* β, introduced by Trevese and Vagnetti [2] and defined by Eq.1.
In the X-ray band, though, the *photon index* Γ is more commonly adopted, with the power-law spectra described as $n(E) \propto E^{-\Gamma}$. Γ is thus related to α by $\Gamma = -\alpha + 1$; this means that the spectral variability parameter is

$$\beta = -\frac{\Delta\Gamma}{\Delta \log F} \,. \tag{2}$$

Unfortunately, the photon index Γ is not available. However, it is related to the *hardness ratio*, and thus it is useful to look for correlations between the hardness ratios and the fluxes of the ensemble set.

3.1. HR-Flux Correlations

Among the many quantities presented, the XMMSSC-DR5 catalogue lists the integrated flux at five different bands, labeled from `EP1_FLUX` to `EP5_FLUX` (that we rename as F_1 to F_5 for simplicity), respectively in the bands 0.2-0.5 keV, 0.5-1 keV, 1-2keV, 2-4.5 keV and 4.5-12 keV. XMMSSC-DR5 also lists *hardness ratios* between fluxes in different bands, defined as follows:

$$HR_i = \frac{CR_{i+1} - CR_i}{CR_{i+1} + CRi} \,, \tag{3}$$

where CR_i is the *countrate* in the band i and is related to the flux F_i by the relation

$$F_i = a_i CR_i \,, \tag{4}$$

with the conversion factors a_i provided by the on-line utility `WebPIMMS`[1].

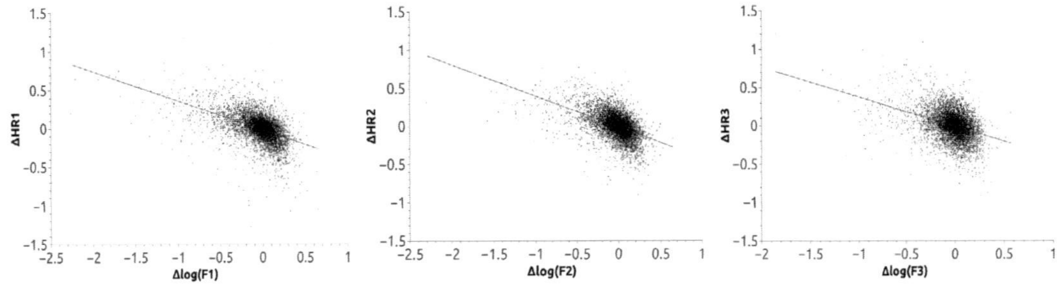

Figure 1. Variations of hardness ratio plotted versus the variations of $\log F$ for the three considered bands. The blue line is the least-squares best fit for a linear trend.

We also had to take into account that different sources have different average fluxes and hardness ratios among each other. Thus, instead of correlations between HR_i and $\log F_i$, we studied the variations from the mean value of each source of hardness ratio, and the variations, also from the mean value of each source, of the flux logarithm.
We have calculated linear fits for $\Delta HR_{1,2,3}$ vs $\Delta\log F_{1,2,3}$, respectively. Table 2 shows the slopes obtained for each linear fit and the correlation coefficient, while Fig.1 shows the three fits in the ΔHR_i-$\Delta\log F_i$ planes. The probabilities of finding these correlations by chance are negligible, given the high number of points involved.

[1] https://heasarc.gsfc.nasa.gov/cgi-bin/Tools/w3pimms/w3pimms.pl

Table 2. Slopes and correlation coefficients obtained from ΔHR vs $\Delta\log F$ linear fits

Band considered	$\frac{\Delta HR_i}{\Delta \log F_i}$	Correlation coefficient r
ΔHR_1 vs $\Delta\log F_1$	-0.380 ± 0.008	-0.508
ΔHR_2 vs $\Delta\log F_2$	-0.409 ± 0.008	-0.507
ΔHR_3 vs $\Delta\log F_3$	-0.388 ± 0.011	-0.384

3.2. β-HR relation

As we hinted earlier on this section, the spectral variability parameter β can be related to the hardness ratio.

$$\frac{\Delta HR}{\Delta \log F} \simeq \frac{dHR}{d\Gamma}\frac{\Delta\Gamma}{\Delta \log F}.$$

Using Eq.2 we obtain that

$$\frac{\Delta HR}{\Delta \log F} \simeq -\beta\frac{dHR}{d\Gamma},$$

which means

$$\beta \simeq -\frac{\frac{\Delta HR}{\Delta \log F}}{\frac{dHR}{d\Gamma}}. \tag{5}$$

4. Data Analysis and Results

In order to estimate the spectral variability parameter β we need to consider Eq.5.

The numerator of such equation can be taken from Table 2. It should be stressed that this is an average ensemble estimate, and that the spectral variability behaviour can be different from source to source.

Given the absence of the photon index Γ from our catalogue, we proceeded with simulated data in order to numerically compute the denominator of Eq.5 as a function of Γ. We have made use of the `X-SPEC` v.12.9.0 software package [12] to generate some spectral models that could fit our typical sources, using the command `model zwabs*powerlaw`, that also takes redshift into account. According to the standard unified model of AGN [13], type-1 AGNs are unobscured, since they are face-on oriented, that means that the obscuring torus does not affect this kind of active galaxies. One of the spectral models we have considered, therefore, is the unobscured power-law model. Although this is the most likely scenario, one cannot exclude that X-ray obscuration processes are at work, indipendently from optical/UV obscuration processes. Thus, we have also studied some models that include a moderate absorption over a power-law background. We have only considered models with $\log N_H$=21 and $\log N_H = 22$. We generated models with values of Γ ranging from 1 to 3, the typical values for AGNs, and redshift z between 0 and 3. For each model we have computed the integrated fluxes in bands 0.2-0.5keV, 0.5-1 keV, 1-2 keV and 2-4.5 keV (or F_1 to F_4 for short), necessary to compute the simulated hardness ratios HR_1, HR_2 and HR_3.

In fact, using Eq.3 and Eq.4, and defining $k_i = 1/a_i$, we can write the hardness ratio as:

$$HR_i = \frac{k_{i+1}F_{i+1} - k_iF_i}{k_{i+1}F_{i+1} + k_iF_i}. \tag{6}$$

We were able to compute therefore, for each model, the values of $HR_{1,2,3}$ and the corresponding values of β, by computing the numerical derivatives of HR over Γ.

For typical values of Γ in the range 1-3, this parameter has negative values (see Fig.2), which means that a *softer when brighter* behaviour is globally present. It is to be stressed that these results extend the trend previously discovered for local sources to a wide redshift and luminosity range.

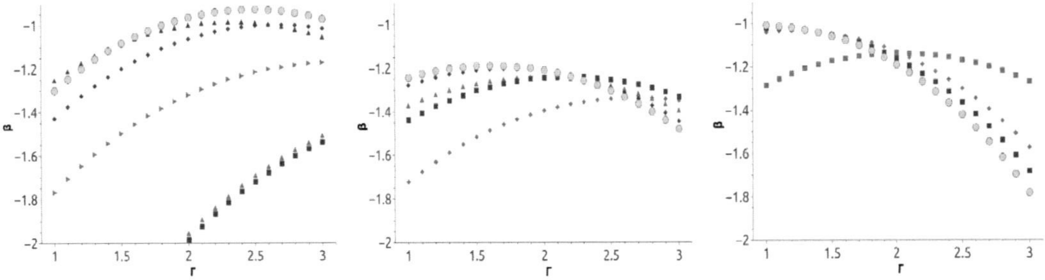

Figure 2. β vs Γ plots for some of the models studied, computed using $HR1$-$F1$, $HR2$-$F2$ and $HR3$-$F3$ correlations, from left to right. The green circles represent the unabsorbed power-law model, while moderate absorbed power-law models are represented by blue ($\log N_H = 21$) and red ($\log N_H = 22$) symbols. Squares, diamonds, up-pointing triangles and right-pointing triangles refer to redshift $z = 0, 1, 2, 3$ respectively.

Aknowledgements
We thank the organizers of the sixth Young Researchers Meeting. We thank the Gran Sasso Science Institute for their hospitality. This research has made use of data obtained from the 3XMM XMM-Newton serendipitous source catalogue compiled by the 10 institutes of the XMM-Newton Survey Science Centre selected by ESA.

References
[1] Sobolewska M A and Papadakis I E, 2009, *MNRAS*, **399**, 1597.
[2] Trevese D and Vagnetti F, 2002, *ApJ*, **564**, 624.
[3] Giveon U, Maoz D, Kaspi S et al., 1999, *MNRAS*, **306**, 637.
[4] Vanden Berk D E, Wilhite B C, Kron R G et al., 2003, *ApJ*, **601**, 692.
[5] Magdziarz P, Blaes O M, Zdziarski A A et al., 1998, *MNRAS*, **301**, 179.
[6] Zdziarski AA , P.Lubinski P, Gilfanov M, Revnivtsev M, 2003, *MNRAS*, **342**, 355.
[7] Ursini F, Petrucci P O, Matt G et al., 2015, `arXiv:1510.01966`.
[8] Paolillo M, Schreier E J, Giacconi R et al., 2004, *ApJ*, **611**, 93.
[9] Schneider D P, Richards GT, Hall P B et al., 2010, *AJ*, **139**, 2360.
[10] Pâris I et al., 2015, in preparation.
[11] Rosen S R, Webb N A, Watson M G et al., 2015, `arXiv:1504.07051`.
[12] Arnaud K A, 1996, *ASP Conf. Series*, **101**, 17.
[13] Antonucci R, 1993, *Ann.Rev.Astron.Astrophys.*, **31**, 473.

Journal of Physics: Conference Series **689** (2016) 012008 doi:10.1088/1742-6596/689/1/012008

Scaling Correlations Amoung Central Massive Objects and Their Host Galaxies

I Tosta e Melo[1], **R Capuzzo-Dolcetta**[1]

[1]Università di Roma "La Sapienza" P.le A. Moro 5, I-00185 Roma, Italy

E-mail: iara.tosta.melo@gmail.com

Abstract. The central regions of galaxies show the presence of super massive black holes and/or very dense stellar clusters. Both such objects seem to follow similar host-galaxy correlations, suggesting that they are members of the same family of Compact Massive Objects. Here we investigate a huge data collection of Compact Massive Objects properties to correlate them with absolute magnitude, velocity dispersion and mass of their host galaxies. We draw also some preliminary astrophysical conclusions.

1. Introduction

The link between the formation and evolution of galaxies and those of their central region is a debated topic. Various studies suggest that massive galaxies, both elliptical and spiral, harbor a SuperMassive Black Hole (SMBH) in their centers, with masses between $10^6 - 10^9$ M$_\odot$. The SMBH masses correlate with various properties of their host galaxies, such as the bulge luminosity [1], mass [2], velocity dispersion [3], and light profile concentration [4].

Galaxies across the entire Hubble sequence also show the presence of massive and compact stellar clusters referred to Nuclear Star Clusters (NSCs). NSCs are bright (4 mag brighter than an ordinary globular cluster), massive ($10^6 - 10^7$ M$_\odot$) and very dense systems, with half light radius of 2-5 pc [5]. In elliptical galaxies, the NSCs are also referred to as Resolved Stellar Nuclei (RSN). All the NSCs contain an old stellar population (age > 1 Gyr), and, in some cases, show the presence of a young stellar population (age < 100 Myr) [6].

The models of formation and evolution of NSCs are still under debate.

Ref. [7] showed that, despite their different morphologies, some galaxies of the Local Group present both a NSC and a SMBH. In such case, the SMBH is surrounded by the NSC. Those objects follow a similar host-galaxy correlation [8], suggesting that they are members of the same family of Compact Massive Objects (CMOs). CMOs constitute a interplay between either SMBH or a compact stellar structure (NSCs, Nuclear Stellar Disk or Resolved Stellar Nuclei).

Ref. [9] found a separation in mass between galaxies that host a SMBH (those with $M > 5 \times 10^9$ M$_\odot$) and galaxies that host a NSC (with $M < 5 \times 10^9$ M$_\odot$). A transition region exists between 10^8 and 10^{10} M$_\odot$ in which NSCs and SMBHs can coexist [10, 11].

A good example is the Milky Way (MW), where a $4\times$ 10^6 M$_\odot$ black hole coexists with an NSC$\sim$4 times more massive ($M_{NSC} \approx 1.5 \times 10^7$ M$_\odot$).

Studies in the literature also showed that the NSC mass versus the host galaxy velocity dispersion (σ) relation is roughly the same observed for SMBHs. Ref. [12] claimed, instead, that the $M_{NSC} - \sigma$ relation is shallower for NSCs ($M_{NSC} \propto \sigma^{1.5}$) than for SMBHs. Moreover, it has been shown that the NSC masses ($M_{NSC} \propto \sigma^{1.5}$) correlate better with the bulge mass, while the SMBH mass displays a tighter correlation with the total mass of the galaxy [13].

Here we investigate these scaling correlations for a set of NSC and SMBH data wider than what already studied in the literature.

The paper is organized as follows: Section 2 presents the data base used for building our sample, and Section 3 describes the methodology. The results and their discussion are presented in Section 4.

2. The database

Ref. [14] selected 43 galaxies of the Fornax Cluster with early-type morphologies (E, S0, SB0, dE, dE,N or dS0,N), using the $F475W$ and $F850LP$ bandpasses of the Hubble Space Telescope (HST) Advanced Camera of Survey (ACS). 31 galaxies out of 43, representing $72 \pm 13\%$ of their sample, are nucleated with the majority of the nuclei bluer than their host galaxies. This approach gives two apparent magnitudes for the nuclei, in the g and z band. The g band values were in this work, because the level of nucleation is slightly larger in this band.

Ref. [15] analysed the nuclei of a sample of 100 early-type galaxies in the Virgo cluster of morphological types E, S0, dE, dEN and dS0. The images were taken with the ACS instrument in Wide Field Channel (WFC) using a combination of the F475W and F850LP filters, roughly equivalent to the g and z bands, in the Sloan Sky Survey photometric system. The authors also concluded that nucleated galaxies are more concentrated toward the center of Virgo cluster. Some nuclei of ACS Virgo Cluster Survey (ACSVCS) are bluer than the underlying galaxies and a central excess is more apparent in the g band rather than in the redder bandpass.

Ref. [16] presented the properties of 228 nuclear stellar cluster in nearby late-type disk galaxies observed with the WFPC2/Hubble Space Telescope (HST/WFPC2), in B and I bands, with distance ≤ 40 Mpc and distance modulus ≤ 33 mag. The authors avoided the most luminous bulges and all ANGs to build the sample, because the presence of strong ANGs would complicate the NSC characterization. They also concluded that largest and brightest NSCs occupy the regime between Ultra Compact Dwarf (UCD) and early-type galaxies.

Ref. [17] analysed the light profile of 200 early-type dwarf galaxies with magnitudes $16.0 < m_{\rm F814W} < 22.6$ mag using the HST/ACS Coma Cluster Survey. NSCs are detected in 80% of the galaxies, thus doubling the sample of HST-observed early-type dwarf galaxies with nuclear star clusters.

We also included in our sample 89 galaxies having a dynamical detection of the central black hole mass. We took information from a catalog presented by [18], and also from a sample presented by [19].

3. Method

Our aim was to estimate CMOs masses for each of the catalogues presented in the previous section of the paper, and compare such values with the mass, velocity dispersion and absolute B magnitude of their host galaxies.

The stellar M/L ratio vs color correlation given by [20] was used to determine the CMOs masses.

The CMOs mass values of ACS Fornax and Virgo Cluster Surveys were obtained by the $g - z$ color correlation. Georgiev's work did not provide the B−V color correlation (reddening corrected), for the HST/WFPC2 archive nuclei. Such values were provided in the Johnson/Cousin photometric system and our reference system is the AB one. We used the correlation given by [21] to do the conversion:

$$\mathrm{B} = \mathrm{B(AB)} + 0.163(\pm 0.004), \tag{1}$$

$$\mathrm{V} = \mathrm{V(AB)} + 0.044(\pm 0.004). \tag{2}$$

The magnitudes for the NSCs given for the Coma Cluster are only in the $F814W$ band, which is equilavent to the I_C band (AB photometry system). The authors did not provide the color-correlation for this sample. We used the CMOs color-correlation average of the previous sample to get our final results.

The galaxy masses for the Fornax and Virgo clusters were computed using the virial theorem:

$$M_\mathrm{gal} = \frac{\beta R_\mathrm{eff} \sigma^2}{G}, \tag{3}$$

where G is the gravitational constant, σ is the velocity dispersion, R_eff is the effective radius of the galaxies, and $\beta = 5$.

For the ACS Fornax Cluster Survey the effective radius values for the galaxies used were provided by [22]. There are no available estimates of the effective radius for FCC 2006, FCC 1340, and FCC 21. The values of the effective radius for galaxies in the ACSVCS were given in their respective catalog.

To compute the galaxy masses in the HST/WFCP2 archive we used the same magnitude vs B−V color correlation for the galaxies and, after the conversion between the reference magnitude systems, from Johnson/Cousin photometric system into the AB system for both bands (E.1 and E.2), we obtained our final results using the stellar M/L ratio vs color correlation formula.

The values for the effective radii are not provided in the ACS Coma Cluster (ACSCCS) catalog. Thus, for it we used the galaxies color-correlation average of the previous sample and we applied such values in stellar M/L ratio vs color correlation formula.

The velocity dispersion values for the galaxies in the Fornax and Virgo Cluster, and those in the HST/WFCP2 archive were taken from the Hyperleda Website[1]; for the Coma Cluster the values are given by [23, 24].

Ref. [14, 16] work also provide the values for the apparent B magnitude for the galaxies in Fornax Cluster and HST/WFCP2 archive, respectively. Such values, for the Virgo Cluster, were taken from [6]. For the Coma Cluster we converted the values of apparent I_C magnitude into apparent B magnitude:

$$F475W - F814W = -0.036(F475W + 18) + 1.13, \tag{4}$$

where $F475W$ and $F814W$ are equivalent to B and I_C bands, respectively.

All the values for the SMBH database were taken in the literature, such as absolute B magnitude [25, 26], velocity dispersion [27], and masses of their host galaxies [25].

4. Comparative discussion of results

Our results are plotted in Figs. 1-4. The existence of evident correlations indicates a direct link among large galactic spatial scales and the much smaller scales of nuclear environment. We have fitted scaling correlations connecting the CMOs mass to various properties of their host galaxies: M_B, σ, and galaxy mass. To obtain the final value of the two fit parameters, a and b, for each galaxy cluster, we applied a linear best fit, i.e. least χ^2 method. Thus the values shown in Table 1 represent the best approximation recoverable.

NSCs are more common in fainter galaxies, with magnitudes between $-21 \leq M_\mathrm{B} \leq -13$, than in brighter galaxies as shown in Figs. 1 and 4. The same behaviour was already predicted by [15].

The presence of NSCs is much more frequent in galaxies with lower velocity dispersion, $\sigma \leq 100$ km/s (Fig. 2).

[1] http://leda.univ-lyon1.fr/

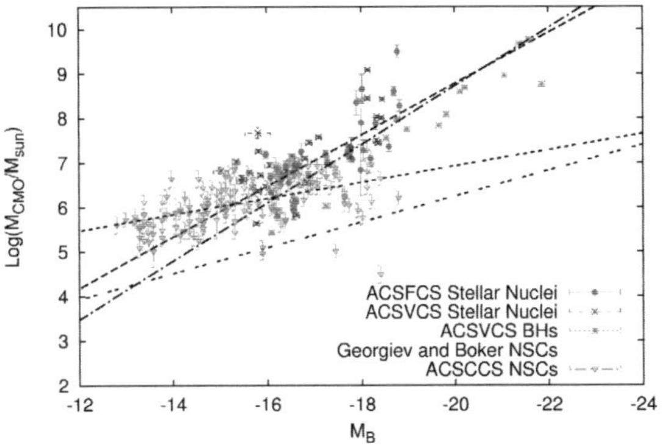

Figure 1. Masses of the CMOs versus the absolute B magnitude of the host galaxies. The black dashed line, the black dashed pointed line, the black two pointed line, and the black pointed line are the least square fit of the Fornax, Virgo, Georgiev and Boker, and Coma sample, respectively.

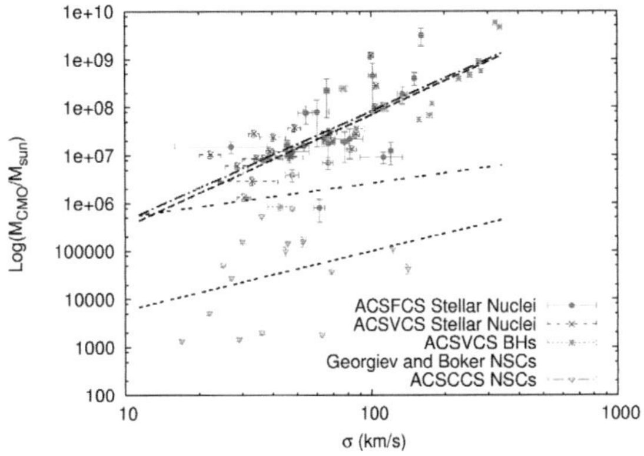

Figure 2. Masses of the CMOs versus the velocity dispersion of the host galaxies. The legenda for the various least square fits is the same of Fig. 1.

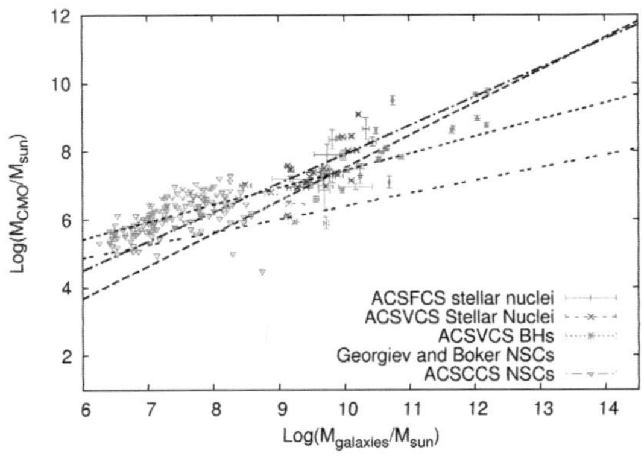

Figure 3. Masses of the CMOs versus the mass of the host galaxies. The legenda for the various least square fits is the same of Fig. 1.

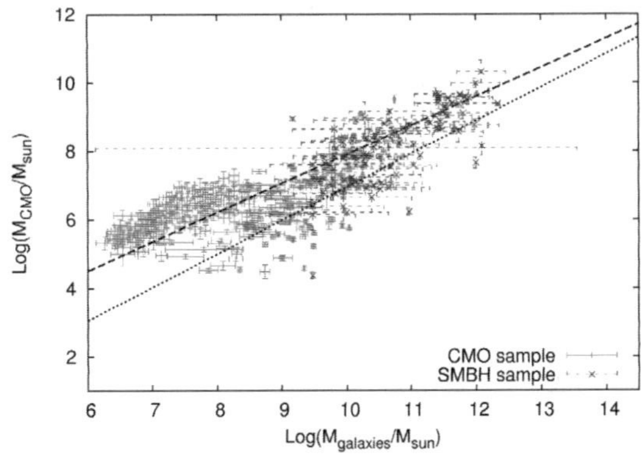

Figure 4. Distribution of the B integrated magnitudes of CMOs for the 571 objects belonging to the ACSFCS, ACSVCS, HST/WFCP2 archive and ACSCCS sample.

Table 1. Valeus of the best least square fits. In the first column numbers 1, 2, 3 and 4 represent the ACSFCS, ACSVCS, HST/WFCP2 and ACSCCS samples respectively. The slope a and linear coefficient b, with their respectives errors for the logarithmic fit of M_{CMO} as function of $M_{\rm B}$, σ, and the mass of the host galaxies are reported in the other columns.

Sample	$M_{\rm B}$ (mag)		σ (km/s)		$M_{\rm galaxies}/M_{\rm sun}$	
	a	b	a	b	a	b
1	-0.57 ± 0.09	-2.71 ± 1.58	1.94 ± 0.81	3.76 ± 1.49	0.95 ± 0.27	-2.07 ± 2.7
2	-0.47 ± 0.04	-1.02 ± 0.81	2.28 ± 0.25	3.35 ± 0.49	0.75 ± 0.09	0.12 ± 0.92
3	-0.28 ± 0.04	0.45 ± 0.89	0.54 ± 0.56	5.17 ± 0.99	0.37 ± 0.11	2.63 ± 1.01
4	-0.14 ± 0.04	3.88 ± 0.66	1.49 ± 0.88	1.64 ± 1.40	0.49 ± 0.06	2.43 ± 0.45

Galaxies with larger masses have more massive CMOs (Fig. 3). It is likely that MBHs and compact stellar structures in these galaxies interacted in the same way during their formation.

5. Conclusions

We studied three different scaling correlations with an updated set of data involving CMOs mass and host galaxies properties. NSCs (or CMOs and SMBH) appear to not to follow the same scaling relation respect to the host galaxy mass, in agreement with the previous studies. Our list of findings are:

(i) Galaxies brighter than $M_B = -18$ host SMBHs, and the existence of such objects in bright galaxies reconcile with the existence, in most of the cases, of an AGN. The lack of NSCs in faint galaxies may be related to the small number of globular clusters in some galaxies, as in the Local Group, where globular clusters are not found in stellar spheroids fainter than $M_B \sim -12.5$.

(ii) We reconfirmed that, with our updated set of data, as one moves to fainter galaxies, the nuclei become the dominant feature while MBHs tend to become less common and, perhaps, entirely disappear at the fainter end [10].

(iii) The M_{NSC} - σ correlation shows a slope between 1.5 and 3, which is in good agreement with that obtained by [1], excluding HST/WFCP2 archive sample.

(iv) It is unclear what exactly the result of any interaction may be: (1) the presence of a single SMBH may evaporate the compact stellar structure, (2) a single or binary SMBH may heat and erode the surrounding compact stellar structure.

Acknowledgments

I. Tosta e Melo acknowledges CAPES-Brazil for support through the grant 9467/13-0.

6. References

[1] Kormendy, J., & Richstone, D. 1995, ARAA, 33, 581
[2] Häring, N., & Rix, H.-W. 2004, APJL, 604, L89
[3] Ferrarese, L., & Merritt, D. 2000, APJL, 539, L9
[4] Graham, A. W., & Spitler, L. R. 2009, MNRAS, 397, 2148
[5] Arca-Sedda, M., & Capuzzo-Dolcetta, R. 2014, MNRAS, 444, 3738
[6] Rossa, J., van der Marel, R. P., Böker, T., et al. 2006, AJ, 132, 1074
[7] Kormendy, J., & McClure, R. D. 1993, AJ, 105, 1793
[8] Ferrarese, L., Côté, P., Dalla Bontà, E., et al. 2006, APJL, 644, L21
[9] Ferrarese, L., Côté, P., Jordán, A., et al. 2006, APJS, 164, 334
[10] Seth, A., Agüeros, M., Lee, D., & Basu-Zych, A. 2008, APJ, 678, 116
[11] Graham, A. W., & Driver, S. P. 2007, MNRAS, 380, L15
[12] Graham, A. W., Erwin, P., Caon, N., & Trujillo, I. 2001, APJL, 563, L11
[13] Erwin, P., & Gadotti, D. A. 2012, Advances in Astronomy, 2012, 946368
[14] Turner, M. L., Côté, P., Ferrarese, L., et al. 2012, APJS, 203, 5
[15] Côté, P., Piatek, S., Ferrarese, L., et al. 2006, APJS, 165, 57
[16] Georgiev, I. Y., Böker, T. 2014, MNRAS, 441, 3570
[17] den Brok, M., Peletier, R. F., Seth, A., et al. 2014, MNRAS, 445, 2385
[18] Savorgnan, G. A. D., & Graham, A. W. 2015, MNRAS, 446, 2330
[19] Ferrarese, L., & Ford, H. 2005, SSR, 116, 523
[20] Bell, E. F., McIntosh, D. H., Katz, N., & Weinberg, M. D. 2003, APJS, 149, 289
[21] Frei, Z., & Gunn, J. E. 1994, AJ, 108, 1476
[22] Ferguson, H. C. 1989, AJ, 98, 367
[23] Matković, A., & Guzmán, R. 2005, MNRAS, 362, 289
[24] Weinzirl, T., Jogee, S., Neistein, E., et al. 2014, MNRAS, 441, 3083
[25] Scott, N., Graham, A. W., & Schombert, J. 2013, APJ, 768, 76
[26] Rusli, S. P., Thomas, J., Saglia, R. P., et al. 2013, AJ, 146, 45
[27] Savorgnan, G. A. D., & Graham, A. W. 2015, MNRAS, 446, 2330

Journal of Physics: Conference Series **689** (2016) 012009 doi:10.1088/1742-6596/689/1/012009

N-body model of magnetic flux tubes reconnecting in the solar atmosphere

L Giovannelli[1], F Berrilli[1], D Del Moro[1], S Scardigli[1], G Consolini[2], M Stangalini[3], F Giannattasio[2], A Caroli[1], F Pucci[1], V Penza[1]

[1]Università di Roma "Tor Vergata", Via della Ricerca Scientifica, I-00133 Roma, Italy
[2]INAF - IAPS, Via del Fosso del Cavaliere 100, I-00133 Roma, Italy
[3]INAF - OAR, Via Frascati 33, I-00040 Monte Porzio Catone (RM), Italy

E-mail: giovannelli@roma2.infn.it

Abstract. The investigation of dynamics of the small scale magnetic field on the Sun photosphere is necessary to understand the physical processes occurring in the higher layers of solar atmosphere due to the magnetic coupling between the photosphere and the corona. We present a simulation able to address these phenomena investigating the statistics of magnetic loops reconnections. The simulation is based on N-body model approach and is divided in two computational layers. We simplify the convection problem, interpreting the larger convective scale, mesogranulation, as the result of the collective interaction of convective downflow of granular scale. The N-body advection model is the base to generate a synthetic time series of nanoflares produced by interacting magnetic loops. The reconnection of magnetic field lines is the result of the advection of the magnetic footpoints following the velocity field generated by the interacting downflows. The model gives a quantitative idea of how much energy is expected to be released by the reconfiguration of magnetic loops in the quiet Sun.

1. Introduction

One of the main goals in solar physics remains the detailed explanation of how the plasma in the higher layers of the Sun has a stable temperature of the order of 10^6 K. Although several possible non-thermal mechanisms have been proved to be able to transport energy from the photosphere, a detailed explanation of the conversion from magnetic energy to thermal and particle energy is still in need of an answer. Besides, a quantitatively and statistically strong observative evidence of these mechanisms is still lacking [1]. Magnetic reconnection events are considered to be one of the main energy source for the heating of the outer layers of the solar atmosphere (chromosphere and corona). Although both the numerical implementation of the magnetohydrodynamics (MHD) theory and the computational power have evolved rapidly in the last years [2], it remains prohibitive to compute full MHD 3D simulations on large temporal and spatial scales and within an environment containing multiple evolving magnetic flux tubes. In order to simulate a computational box with the side of more than 10 Mm on time scales of several hours, we propose a simulation which starts from simplified numerical models able to mimic the behavior of the magnetic loops in the solar atmosphere without needing to solve time consuming MHD equations. Although not solving the proper physical equations of the system, this class of models can provide a synthetic, but realistic, time series of magnetic energy release due to the reconfigurations of magnetic loops.

1.1. Simplified Reconnection Model

In a simplified reconnection model presented in 2006 [3], multiple loops evolve in space and time in a 3D domain, hereafter Hughes Multiple Loops model (HML).

Magnetic loops are represented by semicircumferences on a plane perpendicular to the photospheric plane, and their footpoints, of equal intensity magnetic field and opposite polarity, are anchored on the photosphere. The footpoints move in a random walk, representing the motion of magnetic field advected by the plasma flows. The magnetic energy stored in the loop is proportional to its length and increases as the two footpoints increase their distance. New magnetic loops are injected in the system in order to simulate the magnetic emergence process. Reconnection can occur in the HML model when either two loops collide in 3D space, or two footpoints annihilate. Reconnection is only allowed if it shortens the combined length of the two colliding loops, thus resulting in a release of energy. The model also takes into consideration cascade events in which one reconnection triggers others.

The resulting probability distribution of energy released during reconnections events is well described by a power law with a slope $\alpha_E = 3.0 \pm 0.2$. This value is slightly high compared to the slope obtained by observational data in the UV band of low energy flares $\alpha_E = 1.5 \div 2.6$ [4]. Another difference compared to the flare observations is the absence of any temporal correlation in the waiting time among reconnection events, maybe due to the random nature of the motion of magnetic footpoints.

In order to improve this point and account for the correct spatial and temporal correlations of the process underneath the magnetic footpoints motion we implemented a model similar to the HML. The new model had to include a simplified advection model able to provide a more realistic motion of the footpoints.

1.2. N-body Advection Model

A fluidodynamical simulation [5] has proved that the profile of the vertical and horizontal velocity fields originated by a downflow is well approximated by an exponential law. The N-body Advection (NA) model [6] starts from this result, representing a downflow as a point generating an attractive velocity field profile: $v = Ve^{-\frac{d}{\sigma}}$ at a distance d. Every downflow is a sink in the plasma and is the source of an attractive velocity field directed to the downflow, attracting all the other downflows. All downflows are advectively transported according to the horizontal velocity field generated by all the others downflows.

New downflows are added in random positions on the domain; the intensity of the downflow at $d = 0$, V, is assigned from a normal distribution where $\sigma = 1$ and $\mu = 1$, values are normalized to 1 km/s. The domain of the simulation is square with periodic boundary conditions (toroidal geometry). The V parameter has an exponential decay over time, characterized by a constant decay τ, which is related to the coherence time of granules. If the intensity at zero distance V goes below the value of 0.0001 km/s, then the downflow is removed, representing an evanescent downflow. The spatial scale of the simulation is obtained interpreting 2σ, the typical influence scale of a downflow, as the granular scale. The NA model then performs a check on the distance between the downflows, if it finds $d_{ij} < 1$ the two downflows are merged. The new value for the V parameter is the sum of the old ones: $V_k = V_i + V_j$. Mass conservation is imposed forcing the conservation of the total flow $F = \sum_i V_i$ at every time step, adding downflows randomly distributed on the domain. The space-time scales are set for the smaller granular-scales and they spontaneously self-organize in the mesogranular scale, constituted by more stable and intense downflows. The size of the mesogranular scale is set by the proper values in the $\sigma - F$ parameter space [7]. Once the right spatial scale for menogranulation is set, the NA model reproduce the observed temporal scale for the mesogranular convective structures [8–10]. Thus it is possible to use a solar NA model to impose the correct spatial and temporal behavior to the magnetic footpoints of an HML-like model.

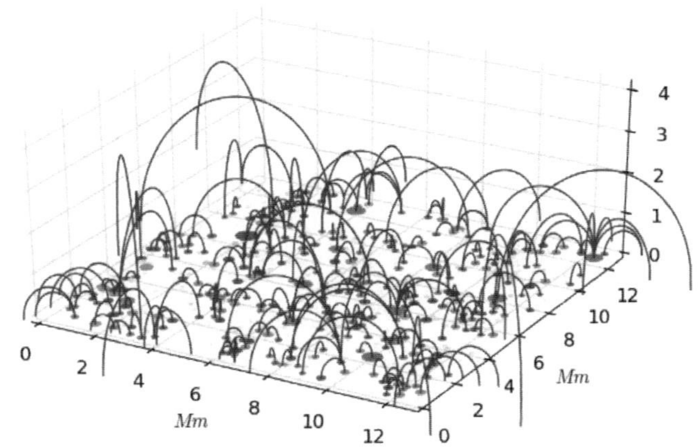

Figure 1. 3D representation of the PASTA simulation with an emergence rate of $6.5 \cdot 10^{12}$ Mx s^{-1} Mm^{-2}. Yellow to red circles: downflows of increasing intensity. Grey circles: magnetic footpoints of positive (darker) and negative (lighter) polarity. In both cases the circle radius is related to the intensity of the downflow or the magnetic field. Blue semicircumferences: magnetic loops.

2. PASTA Model

A model combining the NA and HML models together was proposed [7], introducing two main changes with respect to the HML model:

- the advective velocity field generated by the collective interaction of the downflows is used to move the passive magnetic footpoints with the correct space and time scales.
- the periodic boundary conditions of the NA model were extended to the HML model in order to be able to produce loops with typical scales larger than the computational box.

The model is able to reproduce the power law distribution for the flare released energy over three decades ($10^{23} \div 10^{26}$ erg) with an index $\alpha_E = 2.4 \pm 0.1$, to be consistent with the observations. Furthermore a waiting time distribution is found, which is also in agreement with the observations, following an exponential law with $\alpha = 4.3 \pm 0.3$.

To further implement the model and calibrate the energetic output of the synthetic nanoflares, other observational parameters have to be included in the simulation. We hereby present a new model, the Photospheric Advection Simulation at Tor vergata university ph.d. school of Astronomy (PASTA), which derives directly from [7]. We have rewritten completely the code in python, optimizing the mathematical routines and adding three main features:

- We have considered a distribution of intensity of the magnetic field. Within a magnetic loop the two footpoints have the same strength and opposite polarity, but the field intensity can vary from one loop to the other. The intensity of the magnetic field is extracted from the measured distribution of an exceptional 25 hours long dataset of the HINODE SOT instrument [11].
- Since the intensity of the magnetic field in two interacting loops is not the same, we have introduced a new set of rules in case of interaction, described later in detail.
- The energy of the magnetic loop depends now on the length of the loop and on the intensity of the magnetic field.

2.1. PASTA code description

In our model, multiple loops evolve in space and time in a 3D domain as in [3]. The spatial resolution is set to 1 pixel = 50 km. The footpoints move accordingly to the velocity field generated by the N-body model as in [6], representing the motion of magnetic field passively advected by the plasma.

New loops are added at a fixed rate. When a new loop is added, the position of the first footpoint is random, the second footpoint of the same loop is put at distance d_B in direction ϕ_B. d_B is extracted from a truncated normal distribution. The two linked footpoints have the same intensity for the magnetic field, but opposite polarity. Magnetic footpoints are passively advected by the downflows, following the same spatial exponential law, and so behaving like corks. In Fig. 1 we show one step of the simulation with the 3D loops connecting the magnetic footpoints and the underneath velocity field.

We check for interaction events between footpoints, ruled by the interaction distance set to the spatial resolution of the simulation, 1 pixel, i.e. $50km$:

- CANCELLATION. If two footpoints of the same loop have a distance less than 3 pixels, or a magnetic flux less than 10 G the loop is canceled.
- AGGREGATION. If two footpoints of equal polarity have a distance less than 1 pixel, the two footpoints are moved together in the same position, an average of the former ones. The footpoints are not merged, they remain distinct in order to prevent an arbitrary rule in case of reconnection from a footpoint with different loops departing from it. In this way we have the possibility to simulate a multiple loop footpoint avoiding ambiguity. The footpoints share the same position evolving together.
- ANNIHILATION. If two footpoints of different polarities of two different loops have a distance less than 10 pixels, i.e. the granulation convective scale, there is an annihilation event. These two footpoints are canceled and a new loop is defined between the remaining other ones. We call loop A the loop with the more intense magnetic field and loop B the other. Loop A will interact with loop B as having a magnetic field with the same intensity of loop B. Loop A is schematized as made up of two loops, loop A', with the same intensity of loop B, and loop A", with the remaining magnetic flux. In case of annihilation or reconnection events, only the A' and B loop interact, while the A" loop remains untouched. At the end in case of an annihilation we will a have a C new loop and the A" loop.
- RECONNECTION. If two loops, represented on the photosphere plane as segments, have an intersection on the periodic domain, named the X point, we measure the distance between the two points in the two loops corresponding to the X point. If this distance is less than 1 pixel we perform a reconnection event, where we create two new loops connecting the positive footpoint of the first old loop with the negative footpoint of the second old loop and vice versa. The reconnection is performed only if the initial energy is greater than the final one; this guarantees a positive release of magnetic energy. This latter event is recorded as a synthetic nanoflare.

Multiple events at fixed time are allowed if a reconfigured loop has the possibility to reconnect with another loop after a first reconnection. In this way a flare resulting from a cascade event of simultaneous reconnections is recorded. The possibility of one loop to cut other loops while reconfiguring, and in this way reconnecting with them, is not taken into consideration for simplicity. We do not impose the conservation of the number of loops in the system, although the datasets show a steady number of loops during the simulation.
The energy of a loop is defined integrating the energy density $B^2/8\pi$ over the volume of the magnetic loop:

$$E[\text{erg}] = \frac{\pi}{16}\, d_{loop}\, (B \cdot B_r)^2 \tag{1}$$

where B is the magnetic field intensity and d_{loop} is the distance between the footpoints on the photosphere and thus the diameter of the semicircumference. The diameter of the magnetic flux tube, $2B_r$, is set to the spatial resolution of the simulation, i.e. 50 km.

2.2. Exploring the Magnetic Emergence Rate Parameter

We mainly focus our analysis on the exploration of the magnetic emergence rate parameter. New measures of the emerging magnetic couples have shown that the previous values known in the literature were underestimated at least by an order of magnitude [12, 13]. As we continue increasing spatial resolution and magnetic sensitivity, the small magnetic field contribution becomes very important in the emergence process of magnetic flux tubes. We have investigated the relation between the simulated flare events and the emergence process, trying to estimate the energy released for the observed emergence rate and understand if it is sufficient to explain the coronal heating.

Six simulations are computed, with the same dynamical advection parameters, and increasing magnetic emergence rate, from $7.8 \cdot 10^{11}$ Mx s^{-1} Mm^{-2} to $2.6 \cdot 10^{13}$ Mx s^{-1} Mm^{-2}, doubling the magnetic emergence rate from one simulation run to the other. The aim is to find out the relation between magnetic emergence rate and the probability distribution of the energy released by the synthetic nanoflares.

The domain is a square matrix with a 256 pixels side. The total intensity of the downflows is set to 1000. The spatial scale is set by the $\sigma = 10$ parameter, thus having 20 pixels = 1 Mm. The temporal scale is set by the parameter $\tau = 10$ (simulation time unit), interpreted as the typical coherence time of a granule (10 minutes). The six runs have a goal to simulate 600 minutes.

3. Results

We obtain from each simulation a synthetic nanoflare time sequence and perform a statistical analysis on the released magnetic energy probability distribution function (PDF). PDFs are well described by a power law and the fit is obtained using the method described in [14].

As it is clear from Fig. 2 the index reaches a stable value for a sufficiently high emergence rate of the order of ~ 0.7 Mx s^{-1} Mm^{-2}.

As evident from Fig. 2, the simulation runs relative to the lowest magnetic emergence rate have larger error bars for their power law index. This is an effect of the poorer statistics of flare events in these three runs, due to the lower emergence rate, which gives noisier PDFs. Indeed, the probability to have a reconnection event is a function of the density of the loops, which is not set by any conservation rules, but reaches a steady value non-trivially determined by the magnetic emergence rate parameter. Although the power law index values show an increasing trend as a function of the magnetic emergence rate, taking into account the associated error bars, we can state that all six simulation runs have a power law index compatible with $\alpha_E \simeq 2.4$ (see Fig. 2).

We use the synthetic nanoflare statistics to understand if it is compatible with a plausible mechanism to heat the solar corona. We find that the computed power law index is compatible with the theoretical formulation by Parker, which requires $\alpha > 2$, for a nanoflare energy distribution able to sustain a stable hot corona [15–17].

The computed power law index $\alpha \simeq 2.4$ is also compatible with the index obtained from observations. Solar flares show a similar behavior with a power law index reported $1.5 < \alpha < 2.6$ [4] and evaluated in a reduced range $2.1 < \alpha < 2.6$ using geometrical assumptions [18]. Thus our model has α in the range reported by the observers, although our energy range is between $2 \cdot 10^{20}$ erg $\div$ $3 \cdot 10^{24}$ erg, while the observed energy range is extended at most to the microflare range (see for example the reported power law index $2.3 < \alpha < 2.6$ for the flare in the range 10^{25} erg $\div$ $3 \cdot 10^{26}$ erg from UV observations [19]).

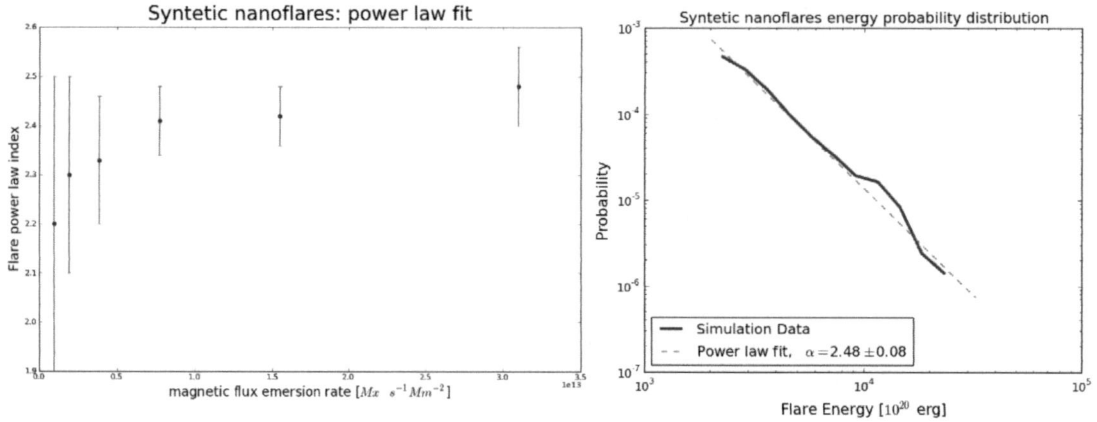

Figure 2. Left: Power law index of the energy probability distribution for synthetic nanoflares as a function of the magnetic flux emergence rate. Right: example of the power law fit for the nanoflare energy PDF in the case of the highest magnetic emergence rate.

4. Summary

The simulation performed in our study combines a model able to mimic the correct spatial and temporal scales of granulation and mesogranulation with a simplified description of the reconnection events between magnetic coronal loops.

The model is able to reproduce the observed dynamical properties of small scale magnetic elements and produce a synthetic nanoflare time sequence over 600 minutes, on a computational domain equivalent to a square with a 12.8 Mm side of the solar phostosphere.

We have explored the magnetic emergence rate parameter of the simulation, introducing a new observational constraint. The nanoflare PDF is well fitted by a power law with an index compatible with $\alpha \simeq 2.4$, which is in agreement with flare observations and compatible with the requirement stated by Parker ($\alpha > 2$) for a nanoflare heated solar corona.

References

[1] De Moortel I & Browning P 2015 *Philosophical Transactions A* **373** 40269
[2] Peter H 2015 *Philosophical Transactions A* **373** 50055
[3] Hughes D Paczuski M Dendy R O Helander P & McClements K G 2003 *Physical Review Letters* **90** 131101
[4] Charbonneau P McIntosh S W Liu H-L & Bogdan T J 2001 *Solar Physics* **203** 321
[5] Rast M P 1998 *Journal of Fluid Mechanics* **369** 125
[6] Rast M P 2003 *The Astrophysical Journal* **597** 1200
[7] Viticchié B Del Moro D & Berrilli F 2006 *The Astrophysical Journal* **652** 1734
[8] Del Moro D 2004 *Astronomy & Astrophysics* **428** 1007
[9] Berrilli F Del Moro D Consolini G *et al* 2004 *Solar Physics* **221** 33
[10] Berrilli F Del Moro D Russo S Consolini G & Straus T 2005 *The Astrophysical Journal* **632** 677
[11] Giannattasio F Del Moro D Berrilli F *et al* 2013 *The Astrophysical Journal Letters* **770** LL36
[12] Martínez González M J & Bellot Rubio L R 2009 *The Astrophysical Journal* **700** 1391
[13] Martínez González M J *et al* 2012 *The Astrophysical Journal* **755** 175
[14] Alstott J Bullmore E & Plenz D 2014 *PLoS ONE* **9** 85777
[15] Parker E N 1988 *The Astrophysical Journal* **330** 474
[16] Vekstein G & Katsukawa Y 2000 *The Astrophysical Journal* **541** 1096
[17] Aschwanden M J Tarbell T D Nightingale R W *et al* 2000 *The Astrophysical Journal* **535** 1047
[18] McIntosh S W & Charbonneau P 2001 *The Astrophysical Journal Letters* **563** L165
[19] Krucker S & Benz A O 1998 *The Astrophysical Journal Letters* **501** L213

Methodological studies on the search for Gravitational Waves and Neutrinos from Type II Supernovae

Claudio Casentini[1,2]

[1] Università degli Studi di Roma "Tor Vergata", Dipartimento di Fisica, Via della Ricerca Scientifica 1, 00133 Roma, Italy
[2] INFN, Sezione di Roma Tor Vergata, Via della Ricerca Scientifica 1, 00133 Roma, Italy

E-mail: claudio.casentini@roma2.infn.it

Abstract. Type II SNe, also called Core-collapse SuperNovae have a neutrino (ν) emission, as confirmed by SN 1987A, and are also potential sources of gravitational waves. Neutrinos and gravitational waves from these sources reach Earth almost contemporaneously and without relevant interaction with stellar matter and interstellar medium. The upcoming advanced gravitational interferometers would be sensitive enough to detect gravitational waves signals from close galactic Core-collapse SuperNovae events. Nevertheless, significant uncertainties on theoretical models of emission remain. A joint search of coincident low energy neutrinos and gravitational waves events from these sources would bring valuable information from the inner core of the collapsing star and would enhance the detection of the so-called Silent SuperNovae. Recently a project for a joint search involving gravitational wave interferometers and neutrino detectors has started. We discuss the benefits of a joint search and the status of the search project.

1. Introduction

The current model of the Universe is like a puzzle with many pieces missing. Photons are the messengers that describe almost all we know about the Universe. We are investigating the sky at all accessible wavelengths of the electromagnetic spectrum. However, a relevant fraction of the Universe should be composed of matter that does not emitting photons, or whose emission is dominated by the cold matter one (i.e. gas or dust).

It is obvious that we need another probe not related to electromagnetic emission. The detection of gravitational waves and neutrinos seems the most promising discovery tool for the future investigation of the Universe, shedding light on physical phenomena that have been theorized only, such as the explosion mechanism of Core-collapse SuperNovae (CCSNe).

2. Scientific case and physical scenario

In this section we provide a quick overview of a CCSN and how this kind of events could generate gravitational signature. Core collapse Supernovae are among the most energetic events observable in the electromagnetic spectrum, since the star irradiates $10^{53}\ erg$ in few seconds. The most brilliant supernovae in the Universe can exceed the brightness that of their host galaxies. In these sources the kiln of the gravitational radiation is the collapsing core and Type II supernovae results from the rapid collapse and violent explosion of a massive star. Stars generate energy by nuclear fusion of elements: this energy together with the degeneracy pressure of electrons balances the force of gravity and prevents the star from collapsing. Massive stars can fuse elements with atomic number greater than hydrogen and helium and result in a core of iron and nickel (see Table 1) [2].

Process	*Main products*	*Temperature (K)*	*Duration (yr)*
Hydrogen burning	He	7×10^7	10^7
Helium burning	C, O	2×10^8	10^6
Carbon burning	Ne, Na, Mg, Al	8×10^8	10^3
Neon burning	O, Mg	1.6×10^9	3
Oxygen burning	Si, S, Ar, Ca	1.8×10^9	0.3
Silicon burning	Ni, Fe	2.5×10^9	$5\ days$

Table 1 Overview of life stages of a massive star.

Therefore, prior to becoming a supernova, an evolved massive star is organized in the manner of an onion, with layers of different elements undergoing fusion. The outermost layer consists of hydrogen, followed by helium, carbon, oxygen, and so forth. Fusion of iron or nickel produces no net energy output leaving the nickel-iron core inert. Due to the lack of energy output outward pressure, equilibrium is broken. A cataclysmic implosion takes place in seconds, during which the core reaches an inward velocity of up to 23% of the speed of light and temperatures up to $\sim10^{11}\ K$. Neutrons and neutrinos are formed via reversed beta-decay, releasing about $10^{46}\ J$ in a ten-second burst. The collapse is halted by neutron degeneracy, causing the implosion to rebound. The energy of this expanding shock wave together with the neutrino heating is sufficient to accelerate the surrounding stellar material up to the escape velocity, forming a supernova explosion. Depending on the initial mass of the star, the remnants of the core give birth to a neutron star or a black hole.

Gravitational emission is expected only if an asymmetrical collapse of the core occurs. During the collapse phase, before the explosion, theoretical models predict strong gravitational associated with a strong neutrino emission from the neutronization experienced by the core. Many mechanisms during the gravitational collapse can ignite the burst of GWs. A non-comprehensive list of models for the emission mechanisms includes [3, 4]:

- Proto Neutron Star (PNS) pulsations;
- Rotational 3D instabilities;
- Rotating core collapse and core bounce;
- Post bounce convection and standing accretion shock instability (SASI).

The signal can last from few milliseconds to few seconds [5], and the energy converted into GWs before the explosion is of the order of $10^{-10} - 10^{-4}\ M_{\odot}c^2$. A typical supernova explosion event should generate a GW signal in the $10^2 - 10^3\ Hz$ frequency band, with a characteristic amplitude that depends on the local amount of energy released in gravitational waves. As discussed in [2], this quantity is given by:

$$h_c \sim 2.7\times10^{-20}\left(\frac{\Delta E_{GW}}{M_{\odot}c^2}\right)^{1/2}\left(\frac{1kHz}{f_c}\right)^{1/2}\left(\frac{10Mpc}{r_0}\right). \quad (1)$$

90% of the energy is conveyed in the neutrino channel. This model is consistent with observations from SN 1987A, where 19 neutrinos from core collapse have been measured [1].

CCSNe and νs are emitted with a time delay ranging from 1 s and 20 s [5] in the inner regions of the star, the core of the proto neutron star (PNS), which cannot be studied electromagnetically, while they uniquely provide information on-the internal mechanisms in place during a CCSN. From neutrinos we could be able to extract information about the equation of state (EOS) of the PNS and from GWs, information on the multi-dimensional dynamics of the collapse, such as deformation of the nucleus. Conflating data from GW and neutrino observatories in a joint analysis allows to find any coincidence between these signals.

Concerning the number of SN events, the expected rate in our Galaxy and the Local Group of galaxies (up to the distances of the order of 300 kpc) is rather low and probably less than 1 event per two decades.

3. Search proposal and strategy

We look for coincident events from GW and Supernova ν experiments in order to shed light on the physics of the core collapse. The search pipeline will be tested on archival ν and GW data before being applied to incoming data from the next generation detectors. The project currently involves several ν experiments, but the approach is general and can be extended to include more observatories.

Fig.1 shows the analysis pipeline setup based on two main stages: the background characterization and the study of the efficiency of detection to incoming signals. The former is performed by analyzing the statistical distribution of accidental coincidences. The latter is pursued by injecting simulated signals into the datasets and checking the ability of the pipeline to retrieve them. The data coming from different communities could be combined in many different ways. In the following we describe the main steps to design a joint analysis between a network of heterogeneous detectors.

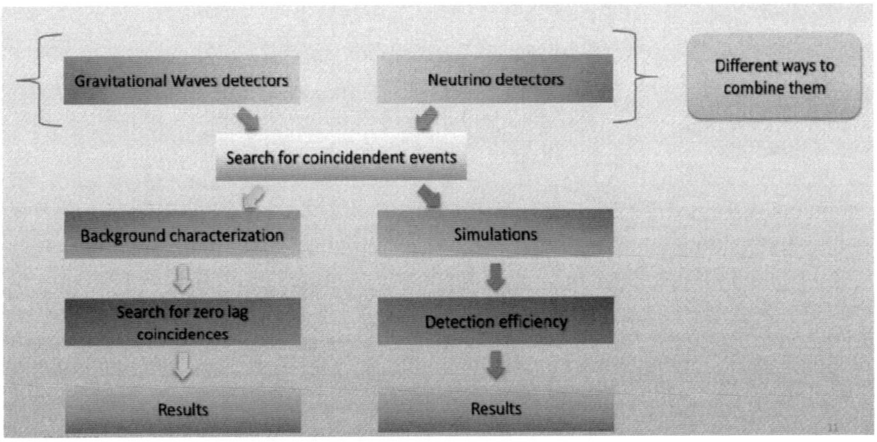

Figure 1 Scheme of the analysis pipeline setup.

One of the key aspects is to account for the single detector duty cycle and hence calculate the common observation time of the network of involved detectors or of the resulting subnetworks. The duty cycle is the percentage of time in which one detector (or a network of detectors) is active. The inactivity periods may be due to several factors, like ordinary maintenance, or any technical/scientific reason.

We need to set a global False Alarm Rate (FAR_{joint}) when searching for coincidences. This value represents a threshold that marks significant events for further investigation. For a Poisson process, this global rate is defined as:

$$FAR_{joint} = (FAR_1 \times FAR_2 \times \ldots \times FAR_N) \times (2t_{coin})^{N-1}, \quad (2)$$

where FAR_i is the selected false alarm rate of the i-th detector and t_{coin} is the time coincidence window considered. The factor 2 accounts for the fact that coincidence searches are generally imposing the relative timing of a GW and neutrino event to be within t_{coin}, but not in a specific order.

The signals that we expect from this kind of sources are burst-like. Due to the fact that this kind of signals are un-modelled and buried inside the detectors background, we need to characterize the statistical distribution of the background in order to estimate the significance of the search and of a candidate GW-neutrino event. To estimate this distribution, we apply unphysical delays between the detectors data streams of the network, in order to remove true astrophysical signals and we search for temporal coincidences. In our analysis we choose to shift the neutrino detector datasets with respect to the GW dataset by a time interval greater than the physical delay between the neutrino and GWs emission. This delay may range between 1s and 20 s. Coincident triggers are selected in the shifted datasets and their distribution describes the number of accidental coincidences expected. The ratio of the number of fake coincidences to the total time of the background collected in the shift analysis represents the rate of accidentals and we study it as a function of a physical quantity of interest (e.g. neutrino energy).

After the background characterization, we look for coincident events in the un-shifted datasets looking for any common signal exceeding the fixed threshold. In the case no results are found, we can set an upper limit on the rate of events assuming a given confidence level [6].

We also have to investigate how well our searching algorithm is able to find a true signal inside the stream of data. We will simulate different GW waveforms and neutrino fluxes from different models and inject them in the GW detectors and in the ν detectors datasets respectively. The detection efficiency will be determined by the percentage of the injected signal found in the output.

4. Involved experiments

In the context of this project, we are working with data from the three gravitational waves detectors of Advanced LIGO (aLIGO) [7] and Advanced Virgo (AdV) [8], and the three neutrino detectors Large Volume Detector (LVD) [9], Borexino [10] and IceCube [11]. A general overview of detectors is provided in the following.

Two main methods for detecting gravitational waves have been implemented in the currently working instruments. One method is to measure changes induced by gravitational waves on the distance between freely moving test masses. The other method is to measure the deformation of large masses at their resonance frequencies. The first idea is realized in laser interferometric detectors, whereas the second idea is implemented in resonant mass detectors.

Both aLIGO and AdV belong to the first group. AdV is currently in upgrading phase, while the aLIGO detectors were recently upgraded and have been taking data since mid-September 2015.

aLIGO is composed of two 4 km long interferometers arms located in a common vacuum tube in Hanford (USA) and in Livingston (USA). The Virgo detector is a 3 km long interferometer (operated by an Italian-French collaboration) located in Cascina (Italy). These detectors are designed to be power-recycled Michelson interferometers. In their first generation design, these detectors are sensitive to h values of the amplitude of $\sim 10^{-22}$. After the upgrade, the sensitivity will improve by one order of magnitude over whole the detection bandwidth. These instruments are expected to start a new GW astronomy era, measuring at least few sources per day.

Several neutrino detectors are sensitive to emissions from galactic CCSN [12].

LVD is a scintillator detector, operated by the Italian Institute of Nuclear Physics (INFN), able to detect both charged current and neutral current interactions. LVD is mainly dedicated to the detection of neutrino bursts from stellar collapses occurring in our galaxy or in the Magellanic Clouds. Its important feature is the possibility to detect neutrinos in several channels, allowing a more careful study of the energy characteristics of neutrino burst from collapsing stars [13, 14].

Borexino is a scintillator detector born to study low energy (sub MeV) solar neutrinos, located in the Laboratori Nazionali del Gran Sasso. The primary aim of the experiment is to make a precise measurement of the ^7Be neutrino flux from the Sun and comparing it to the Standard solar model prediction. It may also be able to detect neutrinos from supernova within our galaxy. For this reason, Borexino is a member of the Supernova Early Warning System (SNEWS) [15].

The IceCube Neutrino Observatory is a neutrino telescope composed of thousands of sensors distributed over a cubic kilometer of volume under the Antarctic ice. It is designed to search for point sources of neutrinos in the TeV range. Despite the fact that individual neutrinos expected from supernovae have energies well below the IceCube energy cutoff, IceCube could detect a local supernova. IceCube is also a member of SNEWS.

5. Benefits of the joint analysis

Searching coincidences of GWs and νs will lead to a deeper understanding of the physics inside star cores. A distant event with low statistical significance in GW could have a neutrinos signature, which would increase the detection confidence in the joint analysis. For galactic silent CCSNe, a coincidence with GW would help constraining the physical models governing the dynamics inside the core. Requiring coincidence with gravitational triggers would allow GW and ν detectors to operate at lower thresholds, relaxing criteria for detection. The guideline is to fix the joint false alarm rate (FAR_{joint}) of the search so that the single searches thresholds can be modified while keeping FAR_{joint} constant. We require a FAR_{joint} of the order of 1/1000 years. In this context, following Equation 2, we can tune our search modifying the thresholds on the single experiments. For example, in order to have the required FAR_{joint}, we can relax the ν threshold to the level of 1 event per day and obtain a GW threshold of the order of 1 event per 3 months. This allow us to include more events in the analysis from both the experiments, that otherwise we would have rejected because they have a low *FAR*. Including more event means improving the sensitivity of the experiment for a certain distance. The improvement for GW and ν searches that we would get are reported in Figure 2. Lowering the ν threshold would correspondingly lead to an increase of the probability to detect neutrinos from distant supernovae and also lead to a 10-20% gain in GW sensitivity (blue solid line with respect to the red dashed line), following [16]:

$$E_{GW} = \frac{\pi^2 c^3}{G} D^2 f_0^2 h_{rss}^2. \tag{3}$$

In Equation 3, D is the distance from the source expressed in kpc, f_0 is the signal frequency and h_{rss} is the signal amplitude at the detector.

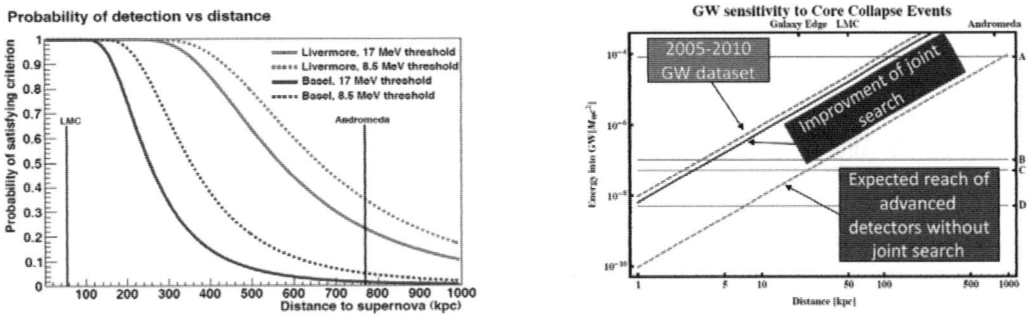

Figure 2 Estimated benefits of the joint search. The left panel shows the gain in probability of detecting two-neutrino events from distant CCSNe as function of the distance, for two different models. The right panel shows the sensitivity to CCSNe by GW detectors as limited by the efficiency of past and foreseen searches. The lines correspond to fixed search sensitivity at the GW detectors, for

narrow band signals at 554 Hz. The horizontal lines represent upper bounds on the energy release for four CCSNe models [3].

Previously studies relevant in the field of GW-ν supernova search are [17, 18].

6. Conclusions

The purpose of this paper is to expose the methodologies in the framework of the search for coincident gravitational wave and neutrino signals from Supernovae. This project is still at an early stage and the main effort focused on creating a common ground where two different communities could interface. For the moment we are re-analyzing archival data in order to set-up the methodologies and preparing the pipeline for new data from advanced interferometers. The future steps will address more in depth the design of the search and its tuning. Furthermore, it is important to characterize the detection efficiency of the network to incoming gravitational waves and neutrinos, by means of simulated signals injected into the data.

Acknowledgements

We thank the organizers of the 6th Young Researcher Meeting and the GSSI for hospitality. This work is done in collaboration with people in several Universities, in particular with Viviana Fafone (Università di Roma Tor Vergata and INFN), Virginia Re (Università di Roma Tor Vergata and INFN), Erik Katsavounidis (MIT), Christian Ott (Caltech), Giulia Pagliaroli (INFN LNGS), Gemma Testera (INFN Genova), Maxim Gromov (Lomonosov Moscow State University), Carlo Vigorito (INFN Torino), Walter Fulgione (INFN Torino) and Lutz Koepke (Mainz University).

References

[1] Hirata K et al. 1987 Phys. Rev. Lett. **58** 1490
[2] Castellani V 1985 Fondamenti di Astrofisica Stellare, Zanichelli Editore
[3] C. D. Ott 2009 Class. Quantum Grav. **26** 063001
[4] Kotake K 2013 C. R. Physique **14** 318-351
[5] Pagliaroli G et al. 2009 Phys. Rev. Lett. **103** 031102
[6] Sutton P J 2009 Class. Quantum Grav. **26** 245007
[7] Abbott B and LIGO Scientific Collaboration 2009 Rept. Prog. Phys **72** 076901
[8] Accadia T et al. 2012 JINST **7** P03012
[9] Cadonati L, Calaprice F P and Chen M C 2002 Astropart. Phys. **16** 361
[10] Agafonova N Y et al. 2008 Astropart. Phys. **28** 516
[11] Halzen F, Jacobsen J E and Zas E 1996 Phys. Rev. D **53** 7359
[12] Scholberg K 2012 Ann. Rev. Nuc. Part. Science **62** 81-103
[13] Giusti P 2014 Pontifical Academy of Sciences Scripta Varia **119** Vatican City
[14] Fulgione W 2006 Mem. S. A. It. Suppl. **9** 388
[15] http://snews.bnl.gov/.
[16] Abbott B and LIGO Scientific Collaboration 2008 Class. Quantum Grav. **25** 039801
[17] Arnaud N et al. 2002 Phys. Rev. D **65** 033010
[18] Arnaud N et al. 2004 Astropart. Phys. **21** 201-221

Journal of Physics: Conference Series **689** (2016) 012011 doi:10.1088/1742-6596/689/1/012011

Search for high-energy neutrinos from GRB130427A with the ANTARES neutrino telescope

Silvia Celli[1,2]

[1] La Sapienza University of Rome and INFN, Piazzale Aldo Moro 5, 00185, Rome, Italy
[2] Gran Sasso Science Institute, Viale Francesco Crispi 7, 67100, L'Aquila, Italy

E-mail: silvia.celli@roma1.infn.it

Abstract. ANTARES is the first deep under-sea high-energy astrophysical neutrino telescope, in operation since 2008, in the Northern Hemisphere. In the light of a multi-messenger approach, one of the most ever intense (photon fluence $F_\gamma \simeq 10^{-3}$ erg/cm^2) and close (redshift $z = 0.34$) transient γ-source, GRB130427A, is considered in the ANTARES physics program for a co-incident search for photons and high-energy neutrinos. The first time-dependent analysis on GRBs neutrino emissions has been performed for this source: Konus-Wind parameters of the γ time-dependent spectrum are used to predict the expected neutrino flux from each peak of the burst, through the numerical calculation code NeuCosmA. An extended maximum likelihood ratio search is performed in order to maximize the discovery probability of prompt neutrinos from the burst: at the end, ANTARES sensitivity to this source is evaluated to be $E^2\phi_\nu \sim 1-10$ GeV/cm^2 in the energy range from 2×10^5 GeV to 2×10^7 GeV.

1. Introduction

The ANTARES neutrino telescope [1] is a three dimensional array of photo-multiplier tubes (PMTs), located at a depth of 2475 m in the Mediterranean Sea, offshore Toulon (France). The detector consists of 12 vertical strings, equipped with 885 PMTs for an instrumented volume of $\simeq 0.02$ km^3. The main scientific goals of ANTARES are the identification of neutrino astrophysical sources and the observation of a neutrino flux exceeding the atmospheric one. Neutrinos are detected through the Cherenkov radiation that is induced by the path in water of ultra-relativistic particles created in a neutrino interaction: track-like signatures are mainly produced by muons from charged-current ν_μ interactions. The PMTs are oriented at 45° downwards in order to maximize the sensitivity to Cherenkov light from upward-going muons: for these events the main background component is constituted of muons induced by atmospheric neutrinos. The typical median resolution achieved for muon tracks is better than 0.4°, allowing neutrino astronomy to be performed.

Neutrinos are unique messengers to study the high-energy Universe as they are neutral and stable, interact weakly and therefore travel directly from the emitting source to the Earth, without being deflected or absorbed. Neutrinos could play an important role in understanding the mechanisms of cosmic rays acceleration and their detection from an astrophysical source would be a direct evidence of the presence of hadronic processes, beyond the well known leptonic ones. Indeed, in the fireball model [2] of Gamma-Ray Bursts (GRBs), the observed electromagnetic radiation is explained as synchrotron emission, followed by inverse Compton

scattering of relativistic shock-accelerated electrons. If protons are also accelerated in the shock outflow, high-energy neutrinos would accompany the electromagnetic signal of the burst [3]. Therefore GRBs are cosmic candidate sources for neutrino telescopes.
In this paper, a time-dependent search is presented: through it, the most intense state of the burst γ emission is selected in order to highly reduce the background and improve the sensitivity by about a factor two with respect to a standard time-integrated point source search.

2. Analysis method

On April 27th 2013, one of the brightest Gamma-Ray Bursts ever detected lit up the high-energy sky: with a photon fluence of the order of 10^{-3} erg/cm^2, GRB130427A turned out to be the strongest burst since 1983. A variability time scale of 0.04 s and a redshift of only 0.34 were measured: assuming the standard ΛCDM cosmological model, this implies an isotropic energy release $E_{\text{iso}} \sim 8.5 \times 10^{53}$ erg. The detection of two high-energy photons, of 95 GeV and 73 GeV [4], soon opened the way to scenarios of hadronic material within the ejecta. Nevertheless, the IceCube Collaboration announced the non-observation of any coincident neutrino signal in their data within ± 1 day around the burst [5].
The GRB130427A light curve exhibited an extremely bright phase, followed by a slow decay of the γ emission: the first pulse alone constituted more than 95% of the burst γ-fluence and was actually released in one twelfth of the whole burst's duration, as can be seen in Fig. 1. The *Konus-Wind* Collaboration analysed the burst in the energy range from 0.02 MeV to 1.2 MeV and provided the time-dependent spectral parameters of each peak, as reported in Table 1 [6]. Since the burst was found to be below ANTARES horizon at the satellite trigger time, a point-like source analysis with data from this neutrino telescope can be performed with up-going events, searching for ν-γ coincident emissions.

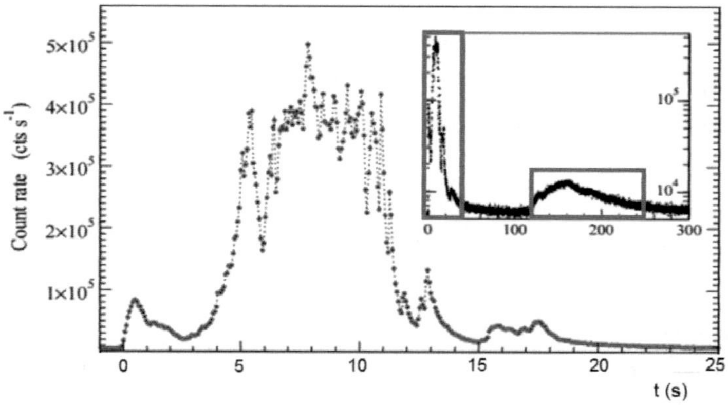

Figure 1. GRB130427A light curve in the energy range from 15 keV to 350 keV: the central panel shows the first peak of the burst, while the upper insert shows the whole burst, with the red boxes representing the two peaks considered in the *Konus-Wind* time-dependent analysis, adopted in this work. (Fig. from [7]).

2.1. High-energy neutrinos from GRB130427A

The theoretical neutrinos flux of the burst is the result of the interaction between a proton, accelerated thanks to the Fermi mechanism, and the observed radiation field. Fluxes from both the first and the second peak of the burst have been calculated through 'Neutrinos from

Table 1. In the table are reported the duration T, the fluence F_γ, the low energy photon index α, the high energy photon index β and the break energy of the spectrum E_γ provided by the γ time-dependent analysis of *Konus-Wind* Collaboration in the energy range from 0.02 MeV to 1.2 MeV.

	T (s)	F_γ (erg/cm^2)	α	β	E_γ (keV)
1st Peak	19	3×10^{-3}	-0.96	-4.17	1028
2nd Peak	130	9×10^{-5}	-1.60	-2.60	240

Cosmic Accelerators' or NeuCosmA code [8]: this is a detailed numerical simulation of the photohadronic interaction processes, accounting for the full proton-photon cross section and including not only the interaction via the Δ^+ resonance, but also kaons and multiple pions production leading to a high-energy component in the neutrino flux. The interaction products and their energy losses are treated individually and the mixing of neutrino flavors on their way to Earth is included. Assumptions were made on the bulk Lorentz factor $\Gamma = 316$, on the jet baryonic load $f_P = 10$ and on the energy fraction going into electrons and magnetic field $e_E = e_B = 0.1$: these parameters can change the neutrino yield for several orders of magnitude, since they affect both the shape and the normalization of the neutrino spectrum [9]. Fig. 2 shows the spectra for track-like events, revealable as muons and antimuons traversing the detector, from each peak: the neutrino spectrum is a broken power law, where the lowest energy break is linked to the break in the photons spectrum, while the higher breaks are due to synchrotron losses of secondary particles. Even in case of neutrinos, the first peak of the burst gives the main contribution. Therefore, the analysis method implemented over it is presented in the following.

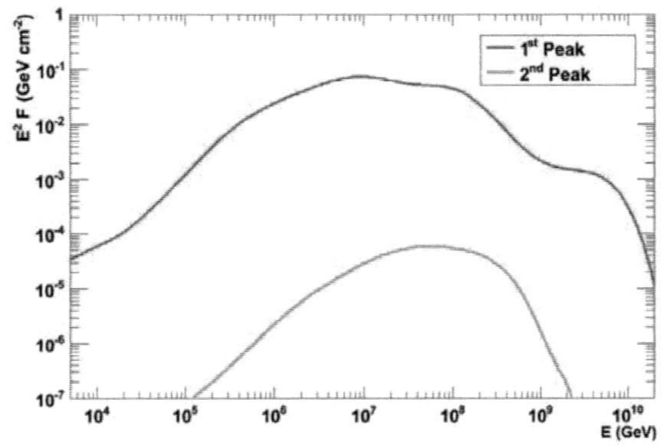

Figure 2. GRB130427A time-dependent NeuCosmA expectations for track-like events: the red curve represents expectations from the first pulse of GRB130427A obtained through *Konus-Wind* γ parameters of Table 1, while the green one represents the second pulse expectations.

2.2. Signal and background Probability Density Functions

For each peak of the burst, signal events due to neutrino interaction according to the expected NeuCosmA fluxes are generated with high statistics through a Run By Run Monte Carlo simulation. These events are then reconstructed in order to compute the detector Point Spread

Function (PSF): the signal Probability Density Function (PDF) is therefore obtained and denoted as $S(\alpha)$, where α represents the space angle between the reconstructed event direction and the GRB's coordinates. In Fig. 3 the cumulative PSF distribution for track-like events is reported: it shows that about 85% of the expected signal events are reconstructed inside an half cone angle of 10°. Since GRBs are transient sources, the angular window of the search can be extended with respect to steady sources search: indeed, the expected background in space-time coincidence with the burst is low due to the small selected time window. Therefore, $\alpha = 10°$ is adopted in the following. The time window is defined to be equal to each peak duration T, where the values for T are shown in Table 1, plus a symmetric extension of 2 s.

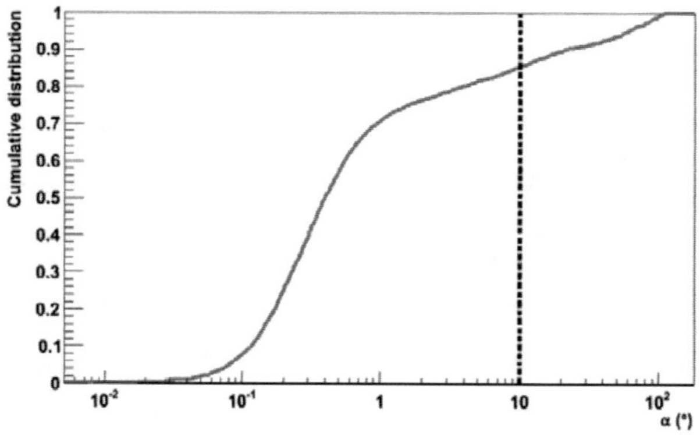

Figure 3. Cumulative distribution of reconstructed events from GRB130427A first peak around its position: the vertical line at $\alpha = 10°$ corresponds to the half cone angle selected for the analysis, since it contains 85% of the expected signal events.

The background PDF $B(\alpha)$ is assumed flat within the defined search cone. In order to estimate the expected mean number of background events μ_b inside the selected search windows, off-time data collected by ANTARES in 2013 are used: the average reconstructed event rate in the direction of the GRB is then corrected in order to take into account the detector efficiency during the run at which the burst triggered.
The track reconstruction algorithm [10] returns the fit quality parameter Λ: a cut on this parameter selects well reconstructed events and is therefore used to optimize the analysis. Both $S(\alpha)$ and $B(\alpha)$ depend on the final choice of the Λ cut.

2.3. Search optimization

Pseudo-experiments are generated in order to simulate n_{tot} signal and background events with space angle α_i, according to the normalised PDFs $S(\alpha)$ and $B(\alpha)$ corresponding to each Λ cut. An extended maximum likelihood ratio [11] is defined as test statistic Q and computed for each pseudo-experiment as:

$$Q = \max_{\mu_s^{'} \in [0;n_{tot}]} \left(\sum_{i=1}^{n_{tot}} \log \frac{\mu_s^{'} S(\alpha_i) + \mu_b B(\alpha_i)}{\mu_b B(\alpha_i)} - \mu_s^{'} \right)$$

where the expected number of background events inside the search windows μ_b is fixed from the previous background evaluation.
The distributions of the test statistic for different numbers of injected events are used to evaluate the Model Discovery Potential (MDP) for a given number of expected signal events μ_s as predicted by the NeuCosmA model. The cut on the quality parameter Λ is then chosen as the

one that maximizes the MDP. Fig. 4(a) shows the MDP as a function of an arbitrary number of expected signal events μ_s, obtained using the PDFs with the optimized cuts for the first peak analysis at 3σ significance.

3. Results

Performing the 3σ optimized analysis presented above on both peaks of GRB130427A, the expected number of signal and background events inside the defined angular and time search windows are $\mu_s = 1.55 \times 10^{-2}$ and $\mu_b = 4.07 \times 10^{-3}$ from the first peak and $\mu_s = 2.87 \times 10^{-6}$ and $\mu_b = 3.54 \times 10^{-2}$ from the second one, with a discovery potential of respectively 2% and 0.3%. It is possible to estimate the ANTARES sensitivity to this source, focusing on the first peak of the emission: in the energy range where 90% of signal events are expected to be detected in ANTARES (from 2×10^5 GeV to 2×10^7 GeV), the sensitivity, evaluated as $E^2\phi_\nu$, is in the range $1 - 10$ GeV/cm^2, a factor 150 above the expected flux, as shown in Fig. 4(b). Even if this burst has been one of the most intense ever, it is not possible to constrain the fireball paradigm with it only: improved results on the detector sensitivity can be obtained through the stacking of more sources, as for example in [12]. Moreover, a km^3-sized detector, like the under-construction neutrino telescopes KM3NeT, will certainly allow the fireball paradigm to be probed in the near future.

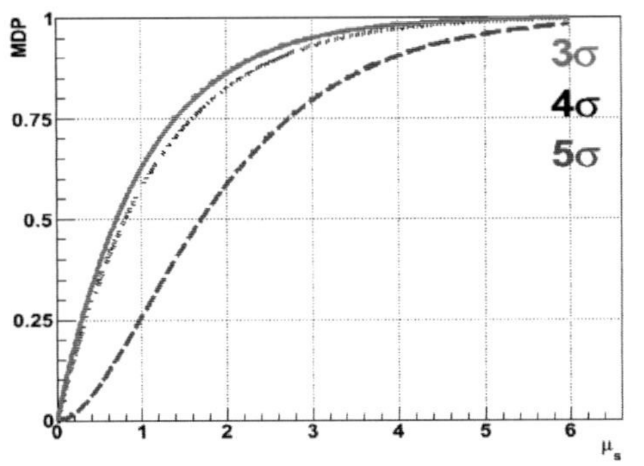

(a)

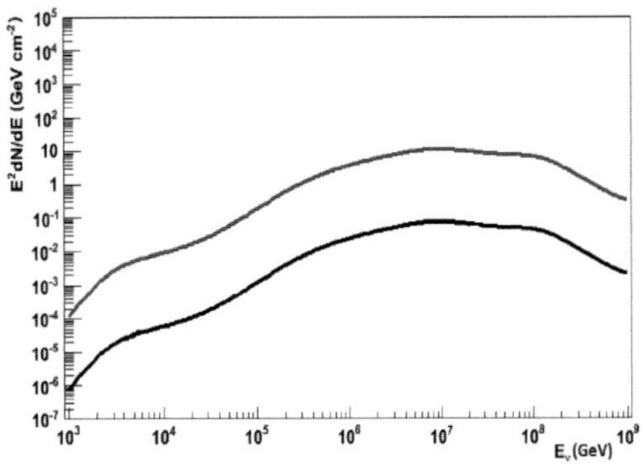

(b)

Figure 4. (a) Model Discovery Potential as a function of the signal flux μ_s for 3σ (red solid), 4σ (black dotted) and 5σ (blue dashed) significance on the first peak of GRB130427A. (b) Expected tracks flux from the first peak of GRB130427A (black) and ANTARES sensitivity to it (blue).

References

[1] Ageron M *et al* (ANTARES Collaboration) 2011 *Nucl. Instr. Meth.* A **656** 11-38
[2] Meszaros P 2006 *Rept. Prog. Phys.* **69** 2259-2322
[3] Waxman E and Bahcall J 1997 *Phys. Rev. Lett.* **78** 2292
[4] Ackermann M *et al* (Fermi LAT Collaboration) 2014 *Science* **343** 42
[5] `http://gcn.gsfc.nasa.gov/gcn3/14520.gcn3`
[6] `http://gcn.gsfc.nasa.gov/gcn3/14487.gcn3`
[7] Maselli A *et al* (Swift Collaboration) 2014 *Science* **343** 48-51
[8] Hümmer S, Baerwald P and Winter W 2012 *Phys. Rev. Lett.* **108** 231101
[9] Guetta D, Hooper D, Alvarez-Muniz J, Halzen F and Reuveni E 2004 *Astropart. Phys.* **20** 429-255
[10] Heijboer A *PhD thesis* 2004 (`http://antares.in2p3.fr/Publications`)
[11] Barlow R 1990 *Nucl. Instr. Meth.* A **297** 496
[12] Adrián-Martínez S *et al* (ANTARES Collaboration) 2013 *Astron. Astrophys.* **559** A9

Journal of Physics: Conference Series **689** (2016) 012012 doi:10.1088/1742-6596/689/1/012012

Study of the high energy Cosmic Rays large scale anisotropies with the ANTARES neutrino telescope

Giulia Illuminati[1]

[1] "La Sapienza" University of Rome and INFN, Piazzale Aldo Moro, 5, 00185 Roma, Italy

E-mail: giulia.illuminati@roma1.infn.it

Abstract. We present the analysis method used to search for an anisotropy in the high energy Cosmic Rays arrival distribution using data collected by the ANTARES telescope. ANTARES is a neutrino detector, where the collected data are dominated by a large background of cosmic ray muons. Therefore, the background data are suitable for high-statistics studies of cosmic rays in the Northern sky. The main challenge for this analysis is accounting for those effects which can mimic an apparent anisotropy in the muon arrival direction: the detector exposure asymmetries, non-uniform time coverage, diurnal and seasonal variation of the atmospheric temperature. Once all these effects have been corrected, a study of the anisotropy profiles along the right ascension can be performed.

1. Introduction

Cosmic Rays (CRs) in the TeV to PeV range are mainly charged particles, believed to originate in our Galaxy, possibly in local astrophysical accelerators such as supernova remnants. After escaping from their sources, CRs propagate through the interstellar medium where they undergo a scattering process due to the presence of magnetic fields. Therefore, they are deflected from their original direction and their trajectories are efficiently isotropized before their arrival at Earth. However, it is predicted that a dipolar anisotropy with an amplitude of $\sim 10^{-3}$ or lower should subsist in their arrival directions. Such a large-scale anisotropy was observed at TeV energies by detectors in the Northern hemisphere, e.g. Tibet ASγ [1], Super-Kamiokande-I [2] and Milagro [3], and then, in the Southern hemisphere, by IceCube [4]. The measured deviation from an isotropic distribution is of the order of 0.1% and the excess region and deficit region have the size of several tens degrees.

2. The ANTARES neutrino telescope

The ANTARES neutrino telescope [5], located in the Northern hemisphere, at a depth of 2475 m in the Mediterranean Sea, about 40 km off the French coast near Toulon, is the first undersea neutrino telescope. Its main purpose is to detect high energy neutrinos from galactic and extragalactic sources [6]. Neutrinos are detected indirectly through the detection of Cherenkov light produced by the path in water of relativistic muons emerging from charged-current muon neutrino interactions in the surroundings of the detector. ANTARES consists of an array of 12 lines anchored to the sea bed and maintained vertical by buoys, connected to a junction box which distributes the electrical power and transmits the data to shore through an electro-cable. Each line supports 25 storeys made of triplets of 10-inch photomultipliers (PMTs) orientated at

Journal of Physics: Conference Series **689** (2016) 012012 doi:10.1088/1742-6596/689/1/012012

45° downwards in order to maximize the sensitivity to Cherenkov light from up-going muons. The PMTs are used to detect the Cherenkov photons which produce a signal whose charge and time information are exploited to recostruct the direction of the muon.

3. Data selection

The data used for this analysis were collected during the period January 2009 to December 2011 with a 12-line detector configuration. The recostruction strategy of the events is based on a maximum likelihood method in which the best set of parameters describing the track is estimated through a recursive fit procedure used to maximise the likelihood function. This function is the probability distribution function of time residuals defined as the difference between theoretical and measured arrival times of the hits on the PMTs.

The following cuts have been applied in order to keep only events that are *suitable* for the analysis. The cut on the zenith angle, $\theta < 78°$, ensures that only down-going events are kept. Only events reconstructed with more than one line have been used since the cylindrical simmetry of the line prevents from obtaining a good reconstruction of the azimuth angle for the single-line events. The cut on the reconstructed energy, $E > 1$ TeV, retains events in the energy range in which the anisotropy has been observed.

4. Data - Montecarlo comparison

The Montecarlo (MC) sample of events used in this analysis is performed with the MUPAGE [7] package which simulates the propagation of the muons produced after the interaction of cosmic rays in the atmosphere. It is a run-by-run simulation: using the measured optical background rates, optical modules conditions and run duration, a realistic simulation of the physics and data taking process for each run is obtained. The comparison between data and simulations is an important step for this analysis: a good agreement between data and MC constitutes a necessary condition to investigate the presence of a possible anisotropy in the arrival distribution of data events compared to those isotropically simulated. Therefore, the first aim of the analysis is to identify the necessary cuts which provide a good agreement between data and MC distributions. The following cuts on the reconondstruction quality parameter, Λ, and on the estimated angular error, β, have been applied: $\Lambda > -6.2$ and $\beta < 1°$. Morever, those events whose track, at the moment of minimum approach to the axis of the detector, is still too far (distance from the detector center on the orizontal plane grater than 50 m and along the vertical axis grater than 200 m) have been rejected. Even if the described cuts have improved the agreement, the simulations are not precise at a level better than a few percent which is not at the required level for anisotropies with amplitudes in the $10^{-4} - 10^{-3}$ range. Therefore, in the analysis method used in this work, the MC distributions are not used to correct real data but the two samples are treated separately and then compared at the end of the analysis.

5. Corrections for spatial asymmetries

An effect which must be accounted for in this analysis is an asymmetry in the ANTARES detector response due to the geometrical configuration of the lines (as shown in Fig. 1 (a)) which results in a variable muon detection efficiency, as a function of the muon direction, as according to the azimuth angle, the number of strings encountered by the muon varies. In order to correct this effect, each event in a certain week j from a given local azimuth bin i was weighted with the ratio $\frac{\bar{n}}{n_i}$ where $\bar{n}$ is the average number of events over the full range of local azimuths in the week j, and n_i is the number of events in local azimuth bin i. This correction is made on a one-dimensional density map of events. This method allows to obtain weights that take into account the fact that the detector is subjected to enviromental conditions which vary over short periods. Fig. 1 (b) shows an example of the azimuth distribution for the week 24/12/2011 - 31/12/2011 with the horizontal red line denoting $\bar{n}$ for that week.

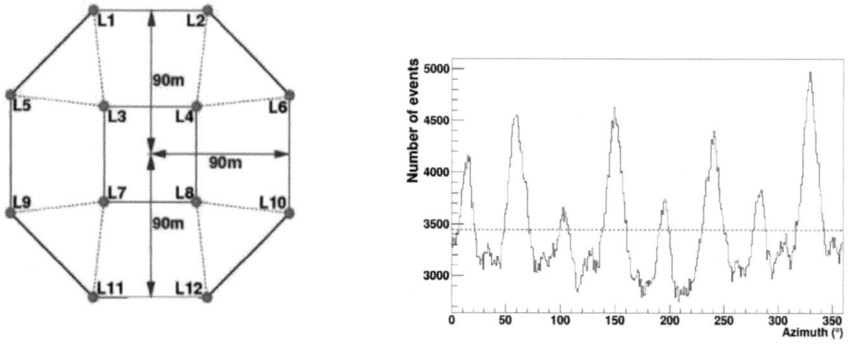

(a) (b)

Figure 1: (a) Lines arrangement in the detector. (b) Azimuth distribution for the period 24/12/2011 - 31/12/2011. The dashed red line shows the value of $\bar{n}$ for the distribution.

In addition to the azimuthal asymmetry a non-uniform cosθ distribution is observed. Fig. 2 shows the cosθ and zenith distributions for the whole sample after the described corrections. It is

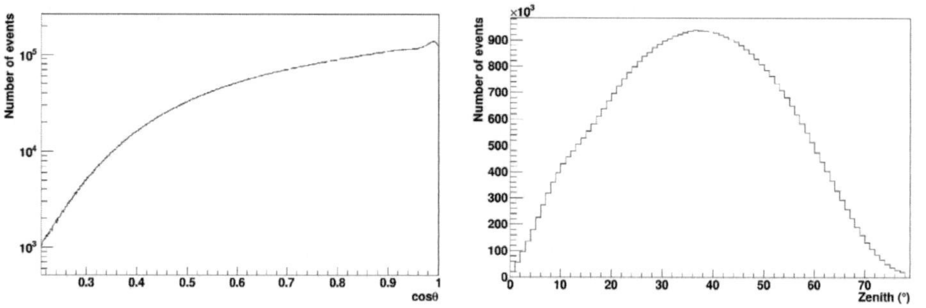

(a) (b)

Figure 2: cosθ (a) and Zenith (b) distribution for the period 01/01/2009 - 31/12/2011 after the corrections for the detector geometrical asymmetry have been applied.

evident that more events arrive from $20° \leq \theta \leq 50°$ compared to those arriving from the zenith and the horizon. This is in contrast with the uniform distribution in cosθ that is expected at the entrance of the atmosphere if the flux of CRs is isotropic and it is due to several effects and selection criteria applied to the secondary muons events. Since it is known that the primary CRs distribution in the upper atmosphere is uniform in cosθ, the idea is to exploit the distribution of secondary muons as an estimate of the overall ANTARES efficiency in detecting CRs by means of downgoing atmospheric muons. Such efficiency can be found normalising the cosθ distribution to its maximum value. The correction for this effect is then obtained by weighting each real and MC event from a given local cosθ bin i by the relative efficiency $w_i = 1/eff(\cos\theta_i)$.

6. Corrections for non-uniform temporal exposure

Another effect which must be accounted for is the non-uniformity in the time coverage of the data: gaps in the detector uptime and uneven run selection due to quality selection introduce non-uniformities into the time coverage of the data which translate into an artificial arrival direction asymmetry in equatorial coordinates. In order to estimate the "exposure" map in equatorial coordinates (Right Ascension, RA, α and declination δ), the following procedure has been used. The duration of each run k in the analysed period has been divided in minutes and, for each minute, every couple of coordinates (ϕ_m,θ_n), in the visible range ($0° \leq \phi < 360°$ and $0° \leq \theta \leq 78°$), has been converted in the corresponding couple of equatorial coordinates (α_i,δ_j) that has then been used to fill a 2-dim histogram $E^k_{i,j} = E^k(\alpha_i, \delta_j)$, with a $6° \times 6°$ binning. The sum over the total amount of runs N of the exposition in each angular bin $\sum_{k=1}^{N} E^k_{i,j}$ gives the coverage map for every couple (α_i, δ_j). Normalising the obtained map to its maximum value provides a new set of weights, functions of the equatorial coordinates, defined as the fraction of the total live time of the detector during the three years in which each angular bin has been visible.

7. Corrections for atmospheric temperature

Another possible effect that can modify the observed muon flux, causing an apparent anisotropy, is the seasonal variation of the atmospheric temperature. Indeed, increases of the air temperature during summer lower the average gas density; the less dense medium allows a longer mean free path of the mesons and increases the fraction of them that decay to produce muons before their first interaction. Since the atmospheric temperature varies over seasonal periods, it is expected to observe a seasonal modulation in the muon flux as well. Such correlation has already been largely observed and studied by experiments like Macro [8], MINOS [9] and Borexino [10].

Since meson decays take place over a range of altitudes where the temperature is changing, it is customary to define an effective atmospheric temperature T, obtained from a weighted average over atmospheric depth [11]. The variation in the observed muon rate R_μ compared to the mean rate over the total data-taking period $< R_\mu >$ can be expressed in terms of a similar change in T as:

$$\frac{R_\mu - < R_\mu >}{< R_\mu >} = \alpha_T \frac{T - < T >}{< T >} \tag{1}$$

where α_T is the "effective temperature coefficient" whose magnitude depends upon the muon energy and hence upon the depth of the detector (α_T is about 0.95 for the ANTARES depth). In this analysis, we have raplaced the muon rate with the ratio between real and simulated events, $R = N_{data}/N_{MC}$, since R shows smaller fluctuations than R_μ. This substitution is valid since possible correlations with the atmospheric temperature are not taken into account in the simulation of MC events. Fig. 3 shows the obtained trend of $\Delta R / < R >$ versus $\Delta T / < T >$, both averaged over periods of 15 days, and the theoretical linear trend.

Temperature fluctuations of the order of $\pm 3\%$ can be observed which lead to an expected R fluctuation of $\sim 2.85\%$ according to eq. 1. However, Fig. 3 shows statistical fluctuations of the ratio between data and MC of the order of 10%. Even if the used data sample with the applied cuts, due to high statistical uncertainties, does not show a correlation with the atmospheric temperature, such correlation is a well known effect already observed by several experiments. Therefore, we have decided to apply the correction due to the atmospheric temperature to the number of observed muons. In particular, each event in a certain 15 days period p has been multiplied by a weight estimated from the averaged temperature in the same period according to eq. 1:

$$w_p = \frac{1}{\alpha_T (\frac{\Delta T}{<T>})_p + 1} \tag{2}$$

8. Analysis of the anisotropy

The corrections described in the previous sections have been adopted in order to obtain the map in equatorial coordinates of the number of observed events as a function of their true

Journal of Physics: Conference Series **689** (2016) 012012 doi:10.1088/1742-6596/689/1/012012

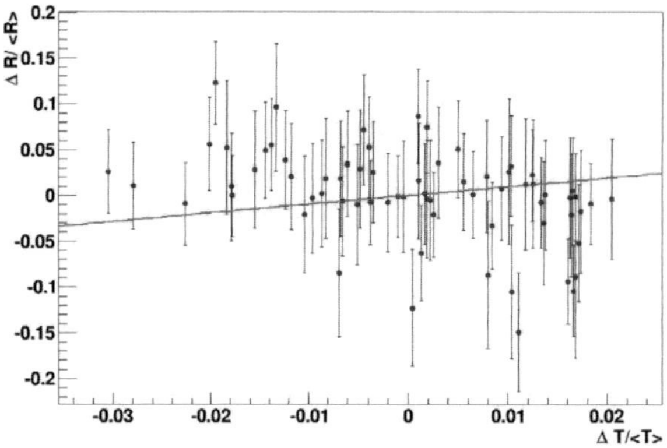

Figure 3: $\Delta R/ < R >$ versus $\Delta T/ < T >$ for the analysed period. The solid red line shows the theoretical linear trend.

arrival direction, so that a possible large scale anisotropy can be verified. In order to study the anisotropy along the RA, we divide the map in equatorial coordinates in 30×60 angular bins ($\Delta\delta = 6°$, $\Delta\alpha = 6°$). For each 6° declination band we evaluate the weighted average of the number of events $\bar{n}_{\delta_j} = \frac{1}{\sum P_{i,j}} \sum_{i=1}^{60} n(\alpha_i, \delta_j) P_{i,j}$. This value is used to normalize to 1 the content of each bin in RA, $n(\alpha_i, \delta_j)$, providing the relative intensity map in which anisotropies are expressed as deviations from 1. Then, the anisotropy profile is obtained as the 1D projection in the RA direction of the 2D relative intensity map. Fig. 4 shows the anisotropy profiles along the RA for data (a) and MC (b). Both trends have been fitted with the equation of a horizontal line and with a first harmonic function in order to test the hypothesis of uniform and dipolar trend. Fig. 4 also shows the result of the fits. The parameters values obtained with the fit are reported in table 1.

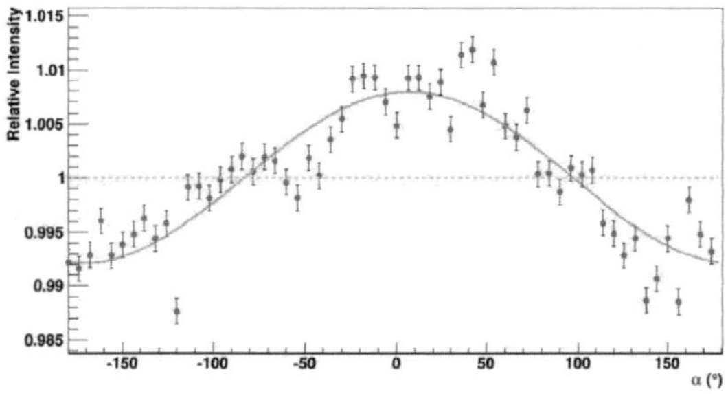

(a)

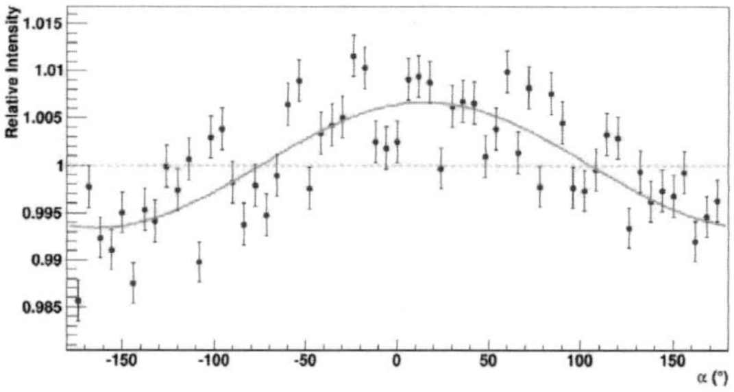

(b)

Figure 4: Anisotropy profiles for data (a) and MC (b). The results of the fit are also shown: the dashed light blue line shows the linear fit ($y = mx + q$ with $m = 0$), the solid green line shows the fit with the first harmonic ($y = A_1 \cos(x - \phi_1) + B$).

Since the distributions show a large dispersion of data, it is unlikely to find a function which represents satisfactorily the obtained trends. The parameters obtained from the fit suggest that for both data and MC the hypothesis of constant value is the worst description of the distributions. Concerning the fit with the harmonic, it is evident that the hypothesis of oscillatory trend cannot be totally excluded for the MC events, whose distributions are expected to be uniform after corrections. This implies that there could still be, in the analysis procedure

Table 1: Columns 1, 2: result from the linear fit $y = mx + q$ with $m = 0$. Columns 3, 4, 5, 6: result from the fit with the first harmonic $y = A_1 \cos(x - \phi_1) + B$.

	q	χ^2/ndf	$A_1(10^{-3})$	$\phi_1(°)$	B	χ^2/ndf
data	1.0000 ± 0.0001	1678/59	7.92 ± 0.22	0.12 ± 0.03	1.0000 ± 0.0001	333/57
MC	1.0000 ± 0.0002	476/59	6.62 ± 0.40	0.26 ± 0.06	1.0000 ± 0.0003	198/57

for example, a systematic effect that causes an apparent anisotropy. However, the fit with the harmonic gives different values of amplitudes and phases for the data distribution with respect to the MC. This suggests that the modulation effect along the RA in the MC distribution is not coincident with the one which causes the modulation in the data distribution.

9. Conclusions

In this work, data collected by the ANTARES detector in the period 01/01/2009 - 31/12/2011 have been analysed with the intention of searching for the large scale anisotropy in the high energy primary Cosmic Rays arrival directions. Since several factors can mimic an apparent anisotropy preventing from obtaining the real distributions of data, the main challenge was to remove such systematic effects with a series of corrections. Then, in order to quantify the scale of the anisotropy in the obtained corrected maps, the RA dependence of the data has been fitted to a first order harmonic function together with a line, used to test the hypothesis of uniform trend for the MC distribution. The obtained results with the harmonic fit suggests that the modulation effect along the RA in the distribution cannot be explained by the same effect which causes the modulation in the MC distribution. The observed "modulation" in RA, that could reflect a higher muon rate in particular seasons of the year, could be related to residual effects due to bioluminescence: higher background could increase the number of "fake" muons but also enlarge the efficiency in real track reconstruction. A larger data sample could give us the possibility to harden the selection cuts and then to remove any residual systematic effects.

References

[1] Amenomori M *et al* 2006 *Science* **314** 439
[2] Guillain G *et al* 2007 *Phys. Rev.* D **75** 062003
[3] Abdo A *et al* 2009 *Astrophys. J.* **698** 2121
[4] Abbasi R *et al* 2010 *Astrophys. J.* **718** L 194
[5] Ageron M *et al* 2011 *Nucl. Instrum. Meth.* A **656** 11
[6] Adrian-Martinez S *et al* 2014 *Astrophys. J. Lett.* **786** L5
[7] Carminati G, Bazzotti M, Margiotta A and Spurio M 2008 *Compututer Physics Communications* **179** 915
[8] Ambrosio M *et al* 1997 *Astropart. Phys.* **7** 109
[9] Adamson P *et al* 2010 *Phys. Rev.* D**81** 012001
[10] Bellini G *et al* 2012 *J. Cosm. Astropart. Phys.* 1205 015
[11] Grashorn E *et al* 2010 *Astropart. Phys.* **33** 140.

Journal of Physics: Conference Series **689** (2016) 012013 doi:10.1088/1742-6596/689/1/012013

Search for cosmic high energy down-going neutrino fluxes from point-like sources with ANTARES

Chiara Perrina

Dipartimento di Fisica dell'Università "La Sapienza" and INFN-Sezione di Roma, P.le Aldo Moro 2, 00185 Roma, Italy

E-mail: chiara.perrina@roma1.infn.it

Abstract. Installed in the Mediterranean Sea, at a depth of $\sim$ 2.5 km, ANTARES is the largest undersea neutrino telescope currently operating. The search for point-like sources with neutrino telescopes is normally limited to a fraction of the sky, due to the selection of events for which the direction of the neutrino candidate has been reconstructed as coming from below the detector horizon, usually referred to as "up-going" events, in order to significantly reduce the atmospheric muons background. In this contribution we demonstrate that through an energy and direction dependent event selection the background can be effectively suppressed so that a part of the region above the horizon can be included in the search. The strategy for the study of a "down-going" neutrino flux is described and the ANTARES sensitivity is presented. No indication of a neutrino signal has been found in the analysed data and upper limits on the flux normalization of a $\propto E_\nu^{-2}$ energy spectrum of neutrinos from several candidate point-like sources in that region have been set.

1. Introduction

Neutrinos represent a valid probe for the exploration of the high energy sky, they provide complementarity with photons and cosmic rays observations since they can emerge from dense media and travel across cosmological distances without being deflected by magnetic fields nor absorbed by ambient matter and radiation. High energy neutrinos (TeV and beyond) are expected to be produced by a wide variety of astrophysical objects. Along with gamma-rays, they originate from the decay of pions and kaons produced in the interactions of accelerated protons and nuclei with matter and radiation inside or in the vicinity of their sources. Candidate neutrino emitters include most known or putative cosmic ray accelerators, ranging from galactic sources such as supernovae remnants or microquasars to the most powerful extragalactic emitters such as active galactic nuclei and gamma-ray bursts [1]. The observation of cosmic neutrinos could allow the identification of known or unknown celestial objects and test the existence of hadronic processes inside astrophysical sources. Cosmic neutrino detectors, usually called "telescopes", need to have a large collection area in order to intercept the faint flux of neutrino events at high energies. This requirement drives to the instrumentation of large volumes of transparent medium (as oceanic water or Antarctic ice) with arrays of photo-sensors, with the aim of detecting the Cherenkov light induced by relativistic charged particles produced when a neutrino interacts with material in or around the detector. Although the identification of muons by ν_μ charged current interactions represents the most straightforward detection channel, showers induced by electron and tau neutrinos can also be detected. Neutrino telescopes are installed at great depths

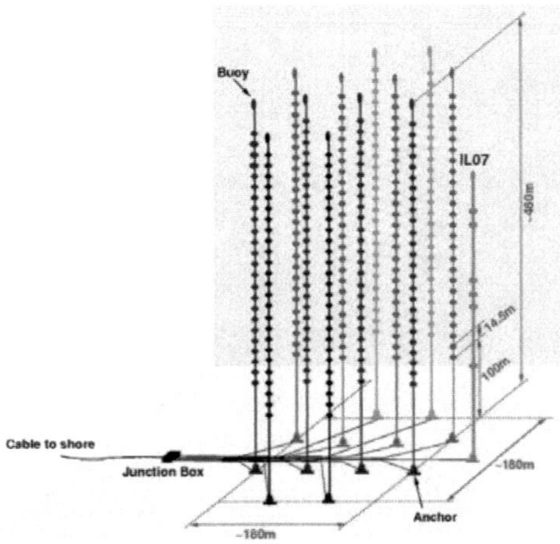

Figure 1: Schematic view of the ANTARES detector [2].

and optimized to detect up-going muons produced by neutrinos that have traversed the Earth, in order to limit the background from down-going atmospheric muons. Atmospheric neutrinos, with an energy spectrum $\propto E_{\nu}^{-3.5}$, traverse the Earth and interact close to the detector representing an irreducible background. Neutrinos of cosmic origin which are expected to have an energy spectrum $\propto E_{\nu}^{-\alpha}$ with $\alpha = 2$ or 2.2 can be identified only on a statistical basis.

2. The ANTARES detector

ANTARES (Astronomy with a Neutrino Telescope and Abyss environmental RESearch) is the first undersea neutrino telescope and the only one currently operating [2]. Its main scientific target is the detection of high energy neutrinos of cosmic origin. It is located on the bottom of the Mediterranean Sea at a depth of $\sim$ 2475 m, in French waters, at $\sim$ 40 km south-east off the coast from Toulon. The apparatus consists in a matrix of 885 PhotoMultiplier Tubes (PMTs) arranged into 12 strings anchored to the sea bed and maintained vertical by buoys, connected to a junction box which distributes the electrical power and transmits the data to shore through an electro-optical cable. The PMTs are orientated at 45° downwards in order to maximize the sensitivity to Cherenkov light from up-going muons. Acoustic devices and inclinometers regularly spread along the strings allow to accurately monitor the position and orientation of the PMTs [3], the time calibration is performed by means of an array of laser and LED beacons [4]. Fig. 1 shows a pictorial illustration of the detector.

3. Data analysis

The analysis presented here has been developed using the data collected by ANTARES between June 2009 and June 2011. The analysed sample has been selected out of data recorded with the detector in perfect operational conditions and corresponds to a total live-time of 366.6 days. As well as for the analysis presented in the last ANTARES letter about the search for point-like sources [5], only events reconstructed as muon tracks have been considered. The special feature lies in the fact that here only down-going tracks have been considered. Fig. 2 and Fig. 3 show the sky visibility for ANTARES in up-going events only and down-going events only, respectively. For both figures, the map on the left panel is drawn in equatorial coordinates, the map on the right panel in galactic coordinates. As it is possible to notice, the region of the sky where declinations are larger than $\sim$ 45° is not covered by the observation of up-going events, but it becomes accessible by studying down-going ones. Therefore, thanks to the analysis described

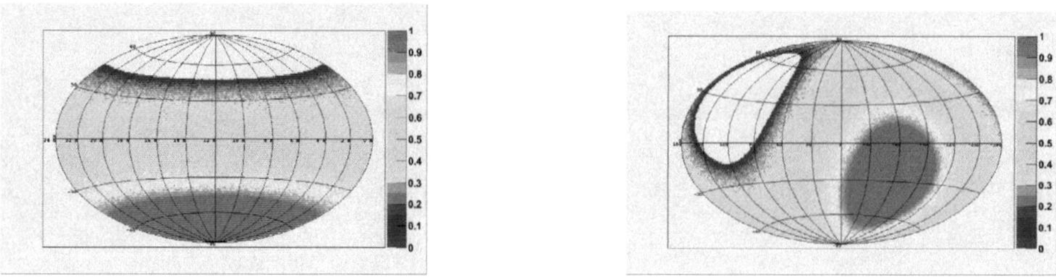

Figure 2: ANTARES sky visibility with up-going events in equatorial coordinates (left) and galactic coordinates (right).

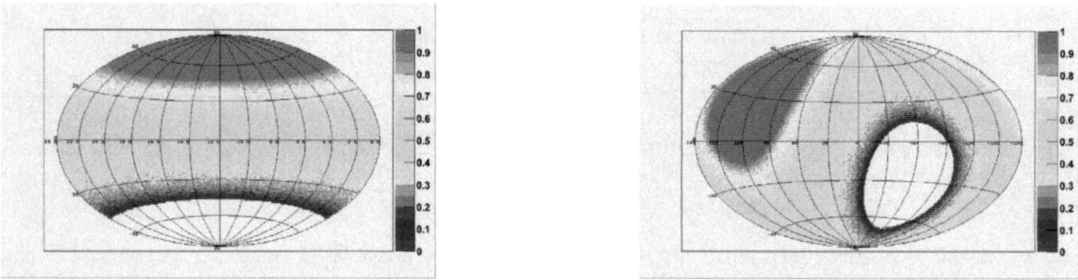

Figure 3: ANTARES sky visibility with down-going events in equatorial coordinates (left) and galactic coordinates (right).

here it has been possible to extend the field of view of ANTARES.

The performed search has been a "candidate-list" one, consisting in looking for an excess of events coming from predefined locations in the sky. Six candidate sources, listed in Table 1, have been selected taking into account their flux in gamma-rays and their visibility for the detector. The big challenge of the analysis of down-going events is due to atmospheric muons which can penetrate through several kilometres of water to the detector, providing the major component of the background. Two characteristics can be used to separate the cosmic signal from the atmospheric background: the difference among the slopes in the energy spectra ($E^{-3.5}$ for the background and E^{-2} or $E^{-2.2}$ for the signal) and the angular distribution, which is isotropic for background but directional for signal from a point-like source. In order to enhance the telescope sensitivity to a cosmic neutrino flux it is necessary to boost the rejection power. An high rejection power has been achieved by means of

- a multivariate analysis, in which both the information about the energy of the event and its incoming direction have been used;
- a statistical technique, called Model Rejection Potential (MRP), which allows to optimize the aperture of the search angular interval around the source position selecting the condition for the best signal to background separation.

A Monte Carlo simulation has been performed in order to estimate neutrino fluxes coming from the sources considered; the background has been estimated from "blinded" experimental data. The "blinding" policy defined by the ANTARES Collaboration is intended to avoid that the selection process and the analysis strategy design are inadvertently tuned on the data sample. The applied procedure consists in the scrambling of the event right ascension: the true right ascension of an event is hidden and a false right ascension is attributed to the event. Muon neutrinos and muon anti-neutrinos charged-current interacted events, coming from the location of the considered candidate sources, with energy ranging from 5 GeV to 100 PeV, have been generated. Fig. 4 shows on the left the bi-dimensional distribution of the point-spread function

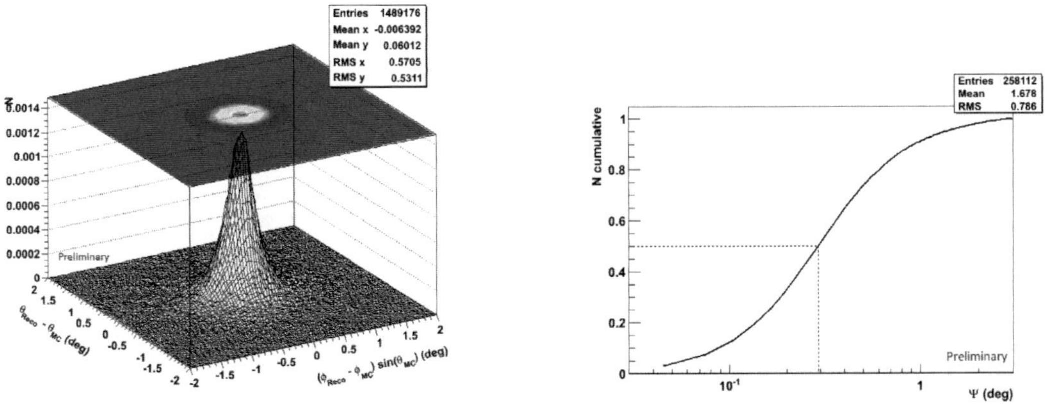

Figure 4: Point-spread function in local coordinates (θ, ϕ) for down-going neutrino events generated from MGRO J2019+37 (left). Angular resolution of down-going neutrino events generated from Tycho, determined as the median of the cumulative distribution of the angle in the space, Ψ, between the generated neutrino direction and the reconstructed muon one. The black-dotted line indicates the median value (right). For both a source spectrum $\propto E_\nu^{-2}$ has been assumed.

for neutrino events generated from MGRO J2019+37. The angular resolution, defined as the median of the cumulative one-dimensional point-spread function distribution, shown in Fig. 4 on the right, is about 0.28°, assuming an energy spectrum $\propto E_\nu^{-2}$.

3.1. Signal to background separation: the multivariate analysis

The analysis of down-going events requests a great effort to obtain the best signal to background discrimination. This purpose has been achieved by performing a multivariate analysis instead of the classic cut-based one. A "Boosted Decision Tree" (BDT) method has been implemented, by using the Toolkit for Multivariate Analysis (TMVA) [6]. Three input variables have been considered: the parameter Λ, a maximum likelihood result of the track fitting procedure, which describes the quality of the track reconstruction [7], the zenith angle θ (ranging from 0° for vertical down-going tracks to 90° for horizontal ones) and the reconstructed energy of the event E_{reco}. Fig. 5 shows the distribution of signal events (by the Monte Carlo simulation) generated from MGRO J2019+37, and background events (by experimental blinded data) for the three input variables. As a result of the analysis of efficiencies and significances, the same cut on the so-called BDT response (> 0.1) has been set for all the sources.

3.2. Signal to background separation: the Model Rejection Potential approach

The aim of this step of the analysis consists in the search for the amplitude of the search cone that maximizes the discovery power. The search cone is defined as a cone with the axis in the direction of the considered source. If a significant excess ($5\,\sigma$) of experimental events over the expected background is observed inside the cone, a discovery can be claimed. The MRP technique [8] has been used. It optimises the amplitude of the cone in order to determine the detector sensitivity (the smallest signal flux that the detector is able to detect) by using signal and background estimates, without the use of experimental data. Around each source direction, cones of different amplitudes (α, from 0.3° to 5° at steps of 0.1°) have been opened and the number of signal events (n_{s}) and background events (n_{b}) occurring in each cone, after the cut in the BDT response, have been determined. According to the MRP method, the strongest constraint on a possible signal flux is obtained when the cone amplitude (α_{best}) minimises the

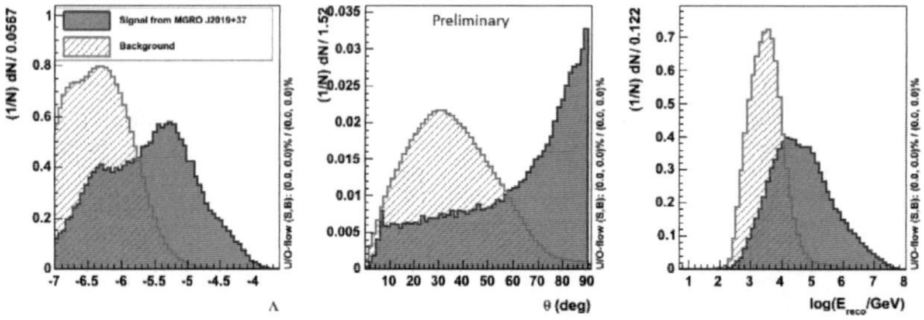

Figure 5: Distribution of training events (signal and background) for the three input variables used for the BDT method: the Λ-parameter, the zenith angle and the reconstructed energy $E_{\rm reco}$ of the event.

so-called Model Rejection Factor (MRF):

$$MRF = \frac{\overline{\mu}_{90}(n_{\rm b})}{n_{\rm s}} \tag{1}$$

where $\overline{\mu}_{90}$ is the Feldman and Cousins 90% confidence interval [9], and its value is:

$$\overline{\Phi}(E)_{90\%} = \Phi_{\rm test} \times MRF_{\rm min} \tag{2}$$

where $MRF_{\rm min} = MRF(\alpha_{\rm best})$. $\Phi_{\rm test}$ represents the assumed neutrino flux from the source, in this analysis $\Phi_{\rm test} = 10^{-8}\ E_\nu^{-2}$ GeV cm^{-2} s^{-1}. In Table 1 the best search cone amplitude for each source is indicated, together with the number of expected background and signal events inside the best cone, finally the ANTARES sensitivity to a neutrino flux, with an energy spectrum $\propto E_\nu^{-2}$, coming from the corresponding source is presented.

4. Results and Conclusions

The final step of the analysis consists in unblinding the experimental data. In such a way it has been possible to know the real incoming direction of detected events, and to count the events occurring within the cone towards each considered source with the aperture determined by means of the MRP technique. No significant excess over the expected background has been found in any of the six considered directions, therefore upper limits have been derived. The most significant cluster, with a significance of $2.1\,\sigma$, has been found for MGRO J2019+37 located at $(\alpha,\, \delta) = (304.65°,\, 36.83°)$. Results are summarized in Table 1: for each source, the number of events that have been observed in the best search cone ($n_{\rm obs}$) is indicated as well as the 90% CL flux upper limit on a muon neutrino flux with an energy spectrum $\propto E_\nu^{-2}$, calculated by using the Feldman and Cousins method [9], is shown. Results are also presented in Fig. 6. Thanks to the analysis of the down-going events, the field of view of ANTARES has been extended at a very interesting region of the sky. The sensitivity of the detector and upper limits for a list of candidate sources with declinations ranging from 37° to 73° have been presented. A worsening of the limits set by the down-going events analysis with respect to those derived by the up-going events analysis was expected. Actually, IceCube [10] too shows a worsening of the sensitivity and of upper limits for neutrino fluxes coming from the Southern Hemisphere ($\sin\delta < 0$) set by means of down-going data with respect to the limits for neutrino fluxes coming from the Northern Hemisphere. Moreover, it should be noticed that the statistics used here is about 30% of the statistics already analysed by ANTARES [5]. By increasing the analysed data sample, even if there will not be discoveries, the limits could go down and ANTARES will be able to put stronger constraint to cosmic neutrino fluxes.

Table 1: The list of candidate sources. For each considered neutrino source, from the second column to the eighth one: the declination of the source, the best search cone amplitude, the number of observed events and expected background and signal events inside the best cone, the sensitivity of the detector and the 90% CL flux upper limit on a muon neutrino flux with an energy spectrum $\propto E_\nu^{-2}$.

Source	δ(deg)	$\alpha_{\rm best}$(deg)	$n_{\rm obs}$	$n_{\rm b}$	n_s	Sensitivity (GeV cm^{-2} s^{-1})	90% CL Upper Limit (GeV cm^{-2} s^{-1})
CTA 1	72.98	0.6	17	21	0.111	2.40×10^{-7}	4.17×10^{-7}
Tycho	64.18	0.6	5	4	0.123	2.36×10^{-7}	4.87×10^{-7}
Cassiopeia A	58.81	0.8	7	4	0.208	1.51×10^{-7}	4.11×10^{-7}
3C 66A	43.04	0.7	3	1	0.215	1.41×10^{-7}	2.99×10^{-7}
3C 254	40.62	0.9	3	4	0.211	1.56×10^{-7}	1.67×10^{-7}
MGRO J2019+37	36.83	0.7	5	2	0.152	1.95×10^{-7}	5.27×10^{-7}

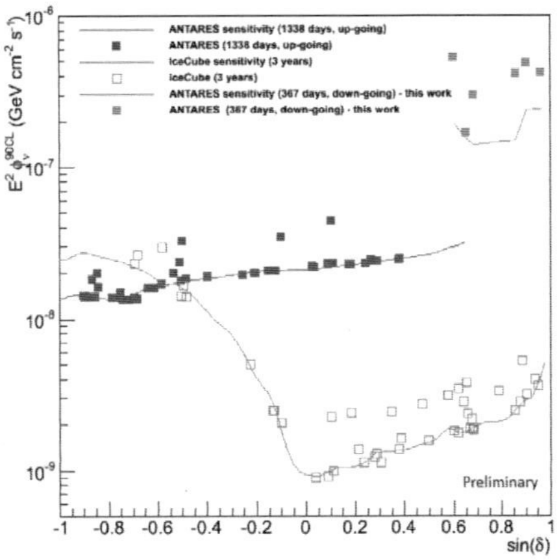

Figure 6: 90% CL flux upper limits on a muon neutrino flux $\propto E_\nu^{-2}$ and ANTARES sensitivity as a function of the source declination. The results of the down-going analysis described here are presented in azure. They provide the telescope sensitivity and upper limits on a cosmic neutrino flux originated from a list of candidate sources with $\delta > 36°$. IceCube and ANTARES results, obtained through the analysis of up-going tracks, are also shown for comparison [5].

References

[1] Becker J K 2008 *Phys. Rept.* **458** 173-246
[2] Ageron M *et al.* (ANTARES Collaboration) 2011 *Nucl. Instrum. Meth.* **A656** 11-38
[3] Adrian-Martinez S *et al.* (ANTARES Collaboration) 2012 *JINST* **7** T08002
[4] Aguilar J A *et al.* (ANTARES Collaboration) 2011 *Astropart. Phys.* **34** 539-549
[5] Adrian-Martinez S *et al.* (ANTARES Collaboration) 2014 *Astrophys. J.* **786** L5
[6] Hoecker A *et al.* 2007 *PoS* **ACAT** 040
[7] Adrian-Martinez S *et al.* (ANTARES Collaboration) 2012 *Astrophys. J.* **760** 53
[8] Hill G C and Rawlins K 2003 *Astropart. Phys.* **19** 393-402
[9] Feldman G J and Cousins R D 1998 *Phys. Rev.* **D57** 3873-3889
[10] Aartsen M G *et al.* (IceCube Collaboration) 2013 *Astrophys. J.* **779** 132

Cross Section Measurements for the ^{23}Na(p,γ)^{24}Mg Reaction at LUNA

Axel Boeltzig[1] **for the LUNA Collaboration**

[1]Gran Sasso Science Institute, Viale Francesco Crispi 7, 67100 L'Aquila, Italy

E-mail: axel.boeltzig@gssi.infn.it

Abstract. LUNA, the Laboratory for Underground Nuclear Astrophysics, is an accelerator facility for measurements of nuclear cross sections of astrophysical interest. The greatly reduced cosmic ray background at LUNA's underground location in the Gran Sasso National Laboratory (LNGS) allows direct measurements of weak reactions at low energies.

One of the reactions currently under study at LUNA is ^{23}Na(p,γ)^{24}Mg, which links the NeNa and MgAl cycles in stellar burning. The LUNA facility is presented, with a focus on the current experimental efforts to study the reaction ^{23}Na(p,γ)^{24}Mg.

1. Cross Sections in Nuclear Astrophysics

Precise cross section data are vital for many considerations in nuclear astrophysics, e. g. in nuclear reaction networks for stellar evolution or nucleosynthesis. Introductions to nuclear astrophysics can be found in [1] and [2].

At astrophysical energies, i. e. energies well below the Coulomb barrier between interacting nuclei, reaction cross sections drop exponentially and become increasingly difficult to measure at lower energies. The combination of the cross section and the energy distribution of the reaction partners (determined by the temperature of the astrophysical scenario) yields the Gamow peak, i. e. the energy region defining the most significant contribution to the reaction rate. Direct measurements in or close to this region are desirable, but often challenging due to the small cross sections involved. The extrapolation of cross section data from higher energies towards the energy region of interest can introduce uncertainties, hence experimental efforts aim at measurements in or close to this region of interest.

In addition to the direct capture cross section described above, the energy dependence of the cross section can be influenced by resonances. Their contribution to the reaction rate depends again on the energy distribution of the particles according to the temperature.

2. Direct Cross Section Measurements at LUNA

The Laboratory for Underground Nuclear Astrophysics (LUNA) is a facility at the Gran Sasso National Laboratory (LNGS), dedicated to direct cross section measurements at low energies. Thanks to its location underground, LUNA benefits from an effective shielding against cosmic rays, which reduces the cosmic ray background from muons by about six orders of magnitude. Muon induced backgrounds (e. g. high-energy gamma rays) are reduced accordingly, which also allows to employ more massive shielding setups against environmental radiation before reaching the limit due to background induced by cosmic rays in the shielding material.

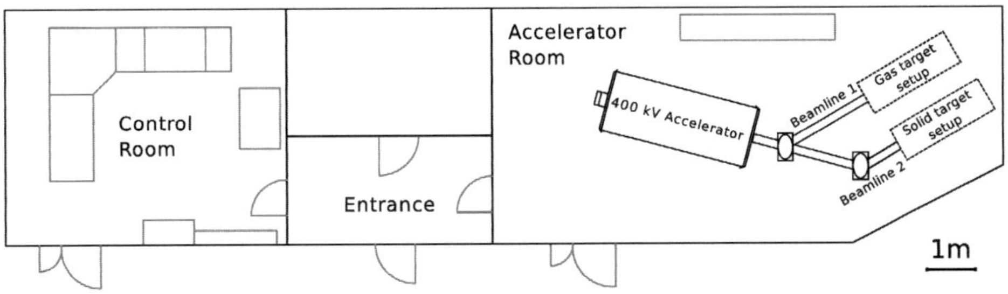

Figure 1. Schematic overview of the LUNA facility underground at LNGS.

The current stage of LUNA employs a 400 kV electrostatic accelerator that can provide beams of protons or ^{4}He with currents of the order of 100 μA. The beam can be directed to either one of the two beamlines located at LUNA, so that two experimental setups can be installed concurrently. Currently, different reactions are under study using a windowless gas target on one beamline and a solid target setup on the other one.

3. The Reaction ^{23}Na(p,γ)^{24}Mg

The ^{23}Na(p,γ)^{24}Mg reaction ($Q \approx 11.7\,\mathrm{MeV}$[1]) provides a link between the NeNa and MgAl cycles in stellar burning, for example in asymptotic giant branch (AGB) stars. For temperatures up to about 10^9 K the rate of this reaction is determined mostly by the direct capture component and three resonances at proton energies of 138 keV, 240 keV and 296 keV respectively (see [3] and references therein).

Whilst the resonance strength is known with relative uncertainties of 18% and 33% for the 240 keV and 296 keV resonance respectively [4], the value of the resonance strength for the 138 keV resonance is largely unknown. The most recent upper limit is given by [5], where the authors also report indications of a signal at 90% confidence level (2.15±1.29 neV). Assuming the resonance strength implied by this indication, the 138 keV resonance would yield the dominant contribution to the ^{23}Na(p,γ)^{24}Mg reaction rate at temperatures around $8{\cdot}10^7$ K. Given the potential role of this resonance in this temperature range, a more precise determination of its strength is sought after, in order to achieve a more accurate determination of the reaction rate.

4. ^{23}Na(p,γ)^{24}Mg Efforts at LUNA

LUNA aims for a measurement of the ^{23}Na(p,γ)^{24}Mg reaction with a solid target setup, in which sodium samples are bombarded by a proton beam. The targets are prepared by evaporation of sodium salts (such as Na_2WO_4) onto a tantalum backing. The area density of the evaporated layer determines the energy loss of the impinging protons in the sodium target layer, defining the "thickness" of the target. Thickness and stoichiometry of the target need to be well known for the analysis of the measured reaction yields. Different techniques are used to study the target composition, in situ (with the target mounted in the setup for bombardment at LUNA) and ex situ (before or after irradiation). Ex situ, Rutherford Backscattering (RBS) and Elastic Recoil Detection Analysis (ERDA) have been used to study the composition of several targets. The main technique used in situ is Nuclear Resonance Reaction Analysis (NRRA): the stronger resonances in ^{23}Na(p,γ)^{24}Mg at 240 keV and 296 keV center of mass energy and a resonance

[1] given by the mass difference of the reaction partners and products: $Q = (m(^{23}\mathrm{Na}) + m(\mathrm{p}) - m(^{24}\mathrm{Mg}))\cdot c^2$

Journal of Physics: Conference Series **689** (2016) 012014 doi:10.1088/1742-6596/689/1/012014

in ^{18}O(p,γ)^{19}F at 151 keV proton energy yield information on the concentration of the two elements in the target. The obtained yields can for example be compared with the expectation from the stoichiometry of Na_2WO_4. The results can also be used to monitor changes in the target composition due to the irradiation.

The challenge for a direct search of the resonance at 138 keV is the detection of event rates that are expected to be as low as a few events per Coulomb of charge accumulated on target. To detect these events with statistical significance, the experimental setup has to provide a large detection efficiency and a low background level in the signal region of interest. To achieve this, the first stage of the experiment employs a segmented Bismuth Germanium Oxide (BGO) detector [6]. The detector provides a coverage close to 4π solid angle around the target and a large gamma ray detection efficiency, due to the properties of BGO, at a moderate energy resolution (4-5% FWHM at 12 MeV). The six identical segments of the detector are read out individually, generating list mode data with time-stamped events. By adding the energy deposition of coincident events, a spectrum resembling that of one large detector can be obtained offline, optimizing the detection efficiency of a signal at the total gamma ray energy released in the reaction. The individual spectra, on the other hand, hold more information on individual gamma rays in the cascades. The full energy peak detection efficiency for a single 12 MeV gamma ray with this detector is about 60%. The efficiency can be lower for a cascade of gamma rays which is often the case for the deexcitation of a level populated in a resonance.

In a second stage, a high purity germanium (HPGe) detector will be used to study the cross section at the high end of the LUNA energy range. A HPGe detector provides a greatly improved energy resolution compared to the BGO detector, at the prize of a lower detection efficiency. Together with its lower solid angle coverage, this detector setup is suited to resolve individual gamma ray energies and determine their branching ratios. Thus the HPGe phase provides complementary information to the measurements with the BGO detector.

The same customized lead shielding is used for both detector setups, providing 10 cm (BGO) or 15 cm (HPGe) of lead around the detector as a shield against environmental gamma rays, whilst allowing fast and easy access to the target chamber at the center of the shield.

Besides environmental backgrounds, beam-induced backgrounds from reactions on nuclides other than the target nuclide are an important issue for the background budget of these direct cross section measurements. If the energy released as gamma rays following a reaction (corresponding to the Q value of this reaction) is larger than in the case of ^{23}Na(p,γ)^{24}Mg, this reaction can directly contribute to the ^{23}Na(p,γ)^{24}Mg region of interest in the BGO detector phase. Reactions with lower Q values can contribute via pile-up. Owing to their large cross sections or strong resonances, certain reactions are particularly critical, e. g. ^{11}B(p,γ)^{12}C ($Q \approx 16.0$ MeV) or ^{7}Li(p,γ)^{8}Be ($Q \approx 17.3$ MeV). Careful selection of materials in the setup is required to minimize contaminations from these elements. With the gamma ray branching information of the involved reactions, the energy deposition patterns in the crystal segments can also be used to discriminate background events.

5. Status and Outlook

Measurements of the ^{23}Na(p,γ)^{24}Mg reaction at LUNA with the BGO detector are ongoing, with a focus on the search for the 138 keV resonance. In addition, measurements of the direct capture component are performed in the accessible energy range. The current limiting factor for the sensitivity of the resonance search are beam induced backgrounds. Several ways to further reduce these backgrounds are under investigation. The HPGe phase will start in early 2016.

A precise model of the background contributions has to be included in the data analysis to extract the ^{23}Na(p,γ)^{24}Mg signal. The analysis of BGO and HPGe data will yield cross sections and, where possible, gamma ray branching ratios. In the overlapping energy region a combined analysis is foreseen to benefit from the advantages of both techniques. Measurements at energies above those at LUNA are planned in cooperation with the Institute for Structure and Nuclear Physics at the University of Notre Dame. A combined analysis of the low- and high-energy data, for example in the framework of R-matrix theory [7], can help to better constrain extrapolations of the cross section data.

References

[1] Rolfs C E and Rodney W S 1988 *Cauldrons in the Cosmos* (Chicago: University of Chicago Press)
[2] Iliadis C 2007 *Nuclear Physics of Stars* 1st ed (Weinheim: Wiley-VCH)
[3] Iliadis C *et al.* 2010 *Nucl. Phys.* A **841** 31–250
[4] Endt P 1990 *Nucl. Phys.* A **521** 1–400
[5] Cesaratto J M *et al.* 2013 *Phys. Rev.* C **88** 065806
[6] Casella C *et al.* 2002 *Nucl. Instr. Meth. Phys. Res.* A **489** 160–169
[7] Descouvemont P and Baye D 2010 *Rep. Prog. Phys.* **73** 036301

A directional Dark Matter argon detector at LNGS

Matteo Cadeddu[1,2]

[1] Dipartimento di Fisica, Università degli Studi di Cagliari, 09042 Monserrato, Italy
[2] INFN - Sezione di Cagliari, Italy

E-mail: matteo.cadeddu@ca.infn.it

Abstract. The potentialities of an argon Dark Matter detector with real-time directional information located at Laboratori Nazionali del Gran Sasso are discussed. Columnar recombination combined with a dual-phase Liquid Argon Time Projection Chamber provides a concrete approach for a directional detector. Even in the case of a minimal angular resolution, a very clear signature of WIMP scattering on target nuclei is shown to be expected. Indeed, the ratio of horizontal to vertical events changes by a factor of four during the day and has a characteristic sidereal-day periodicity.

1. Introduction

At present, a plausible explanation of the nature of Dark Matter (DM) is that it is constituted by yet undetected Weakly Interacting Massive Particles (WIMPs). Nowadays, it is believed that every visible galaxy is surrounded by a roughly spherical halo of DM, extending far beyond the visible disk. The standard non-directional direct detection, which exploits the elastic scattering of WIMPs on nuclear targets in the detector, is experimentally very challenging in particular in what concerns background rejection. An important additional information that could strengthen the DM signature hypothesis is a possible annual modulation of the interaction rate due to the motion of the Earth around the Sun. Indeed, the solar system orbits around the center of the Galaxy, with a velocity $V_{SG} \simeq 220$ km/s. Earth moves around the Sun with a velocity $V_{ES} \simeq 30$ km/s in an orbit inclined about 60° relative to the disk of the Galaxy. The combination of these motions results in a cosine dependence of the count rate with time. Unfortunately, this modulation is quite small and it is very hard to disentangle from other effects, leaving doubts on the authenticity of the few DM observation claims [1]. In addition, numerous radioactive backgrounds are known to exhibit a modulation over the course of the year.

If detectors were able to distinguish the recoil direction of the scattered nucleus after a WIMP interaction, then the signal would have a large and characteristic angular dependence within a sidereal day[1]. Indeed, due to the Earth's rotation around its axis, the detector would face a change in the direction of the WIMP wind and accordingly the angular distribution of the recoil nuclei will vary with time. This modulation in arrival direction produces a much more sizeable and unambiguous signature with respect to a non-directional annual modulation. In principle, assuming the most advantageous experimental conditions, the effect of this modulation can be as large as 100%. The most attractive feature of this kind of modulation is that no known background is correlated with the sidereal period. Actually, a sidereal rate modulation goes out

[1] The sidereal day is about four minutes shorter than the solar day. It correspond to $23^{\mathrm{h}}\ 56^{\mathrm{m}}\ 04^{\mathrm{s}}.090$.

of phase with the solar time, avoiding possible accidental day/night effects which could take place in the laboratory.

Nowadays, a detector capable of measuring the direction of the scattered nucleus has not yet been developed. There are few prototypes [2] of detectors capable of reconstructing the tracks of the recoil nucleus but, since they exploit the usage of rarefied gases, this technology clashes with the necessity of building multi-tonne detectors. For this reason the most promising approach is to explore a detector based on the so-called Columnar Recombination (CR) effect on a Noble Liquid [3], that despite a far from perfect angular resolution can reach large target masses.

2. A directional LAr TPC

By filling a gap in the literature, the sensitivity of an argon detector to the galactic WIMP directional signal has been studied. In particular, a dual phase Liquid Argon Time Projection Chamber (LAr TPC) has been considered, where recoiling nuclei cause argon excitation and ionization. Each event is then detected in two ways: the prompt scintillation light, S1, from argon de-excitation, and the signal from free ionization electrons. The latter drift towards the top of the TPC, thanks to an electric field, and extracted into the argon gas phase, confined in the so-called "gas pocket", close to the top array of photosensors. Accelerated electrons produce a secondary light pulse, S2. A fraction of free electrons, however, recombine with ions in the ion-electron cloud produced by the ionizing track itself.

CR models suggest that the magnitude of the recombination effect should vary with the angle between the field and the track direction. Namely, when the ionizing track is parallel to the electric field the electrons have to drift in a "column" of ion-electron pairs produced by the ionizing track, maximizing the recombination probability. When it is perpendicular, the "column" needs to be crossed and the recombination probability is minimized. The net effect of the electron-ion recombination is the reduction of the ionization signal (S2), and the consequent enhancing of the primary scintillation pulse (S1). An accurate measurement of these two signals and their combination (the sum S1+S2 and the ratio S1/S2) may provide informations on the nuclear track direction and thus on the WIMP directionality. The results shown below were obtained considering an Earth-bound detector, located at Laboratori Nazionali del Gran Sasso (LNGS), with a 100 tonne target mass of argon.

3. Distribution of observables

In a given reference frame, $\mathbf{v_i}$ is the the velocity of the incoming WIMP of mass m_W, $\mathbf{u}$ the velocity of the recoiling nucleus of mass M_N, $\mathbf{q} = M_N\mathbf{u}$ and $E_r = q^2/(2M_N)$ the corresponding momentum and energy, respectively. The azimuthal and zenith angles of the recoiling nucleus are ϕ_r and θ_r, while $\cos\vartheta = \hat{\mathbf{v}}_i \cdot \hat{\mathbf{q}}$ is the cosine of the angle between the incoming WIMP direction and the recoiling nucleus.

For a spin-independent interaction with equal couplings for neutrons and protons the WIMP-nucleus cross section, σ_{W-N}, can be expressed in terms of WIMP-nucleon cross section, σ_n as $\sigma_{W-N}/\mu_N^2 = A^2\sigma_n/\mu_n^2$, where A is the atomic mass and μ_N and μ_n are the WIMP-nucleus and the WIMP-nucleon reduced masses, respectively. The finite size of the nucleus is taken into account thanks to the standard Helm form-factor, $F(q)$.

In the frame where the target nucleus is at rest, the double differential cross section becomes dependent on $\cos\vartheta$; the azimuthal symmetry of the scattering around the WIMP arrival direction gives $d\Omega = 2\pi\, d\cos\vartheta$, thus

$$\frac{d^2\sigma(q, \hat{\mathbf{v}}_i \cdot \hat{\mathbf{q}})}{dq^2 d\Omega} = \frac{d^2\sigma(q, \cos\vartheta)}{2M_N dE_r 2\pi d\cos\vartheta} = \frac{\sigma_{W-N}}{8\pi\mu_N^2 v_i} F^2(q)\, \delta\left(\mathbf{v}_i \cdot \hat{\mathbf{q}} - \frac{q}{2\mu_N}\right). \quad (1)$$

Given a normalized velocity distribution for the incoming WIMP, $f(\mathbf{v}_i)$, and a WIMP mass density, ρ, the double differential recoil rate per unit mass as a function of the nuclear recoil energy, E_r, and of the recoil direction $\hat{\mathbf{q}}$ is

$$\frac{d^2R(E_r,\hat{\mathbf{q}})}{dE_r d\Omega_r} = \frac{\rho}{m_W}\frac{\sigma_{W-N}F^2(q)}{4\pi\mu_N^2}\hat{f}\left(v_{min},\hat{\mathbf{q}}\right) , \tag{2}$$

where $v_{min} = q/(2\mu_N) = \sqrt{2M_N E_r}/(2\mu_N)$ is the minimal WIMP velocity that can give momentum q or energy E_r to the recoiling nucleus and $\hat{f}\left(v_{min},\hat{\mathbf{q}}\right)$ is the 3-dimensional Radon transform of the velocity distribution $f(\mathbf{v})$.

In this work, the Standard Halo Model, i.e. an Isotropic Maxwell-Boltzmann (IMB) WIMP velocity distribution of width σ_v in a inertial reference frame at rest with respect to the Galactic center, is assumed. The corresponding Radon transform [4] is

$$\hat{f}\left(v_{min},\hat{\mathbf{q}}\right) = \frac{1}{\sqrt{2\pi\sigma_v^2}}\exp\left[-\frac{1}{2}\left(\frac{v_{min}+\hat{\mathbf{q}}\cdot\mathbf{V}}{\sigma_v}\right)^2\right] . \tag{3}$$

Therefore, if recoils are measured in a Galactic rest frame, $\mathbf{V} = 0$ and the rate is isotropic. On the other hand, when measured in a frame at rest with respect to the Sun, $\mathbf{V}$ is the Sun velocity relative to the galactic center, $\mathbf{V}_{SG}$, which points towards the galactic coordinates (ℓ_c=90°, b_c=0°), roughly in the direction of the Cygnus constellation. It has magnitude $V_{SG} \approx v_0$, where v_0 is the Galactic orbital speed at the Sun position. For an Earth-bound laboratory the velocity $\mathbf{V}$ can be decomposed as $\mathbf{V} = \mathbf{V}_{SG} + \mathbf{V}_{ES}$, where $\mathbf{V}_{ES}$ is the Earth velocity relative to the Sun.

Given E_{th} and E_{max} the threshold and maximum detector energy, respectively, such that the detector energy range is $E_{th} < E_r < E_{max}$, the direction dependent recoil rate per unit mass is obtained by substituting the Radon transform introduced in Eq. (3) inside Eq. (2) and integrating over the energy range

$$\frac{dR(E_{th},E_{max},\hat{\mathbf{q}})}{d\Omega_r} = \int_{E_{th}}^{E_{max}} dE_r \frac{d^2R(E_r,\hat{\mathbf{q}})}{dE_r d\Omega_r} . \tag{4}$$

In general, a CR directional detector should be able to discriminate only between tracks with different inclination relative to the vertical (electric field) direction. For this reason, the only measurable angle is the polar angle θ_r, and the azimuthal angle should be integrated out. The relevant recoil rate is

$$\frac{dR(E_{th},E_{max},\cos\theta_r)}{d\cos\theta_r} = \int_0^{2\pi} d\phi_r \int_{E_{th}}^{E_{max}} dE_r \frac{d^2R(E_r,\hat{\mathbf{q}})}{dE_r d\Omega_r} . \tag{5}$$

In addition, since a detector based on CR cannot distinguish signals from recoil tracks differing by 180°, i.e. events that differ by 180° are summed together, the relevant rate is the so-called "folded" angular recoil rate

$$\frac{dR_F}{d\cos\theta}(|\cos\theta_r|) \equiv \frac{dR}{d\cos\theta_r}(\cos\theta_r) + \frac{dR}{d\cos\theta_r}(-\cos\theta_r) , \tag{6}$$

which depends only on $|\cos\theta_r|$. In this work the results are shown using the reference values $m_W = 200$ GeV, $M_N = 0.923\,A$, where $A = 39.9$ is the argon atomic mass, $\rho = 0.3$ GeV cm^{-3}, and the IMB $\sigma_v = v_0/\sqrt{2}$. Rates are shown for a reference cross section $\sigma_n = 10^{-46}$ cm^2, which is one order of magnitude smaller than the most stringent limit set by the LUX collaboration [5], and for a recoil energies range from $E_{th} = 50$ keV to $E_{max} = 200$ keV.

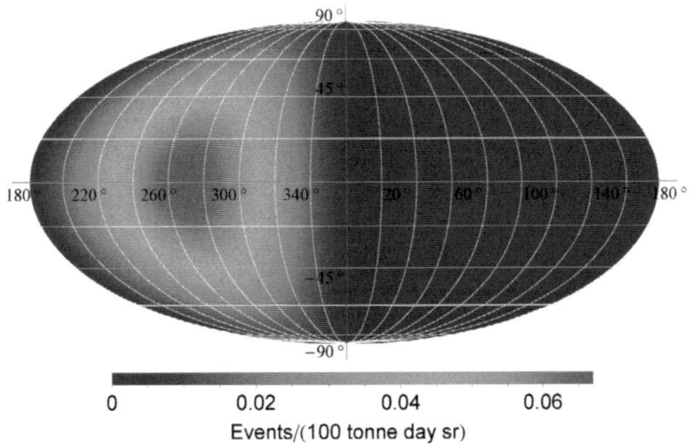

Figure 1. Directional differential recoil rate in argon integrated over the energy range 50÷200 keV, plotted in Mollweide equal area projection map of the celestial sphere in Galactic coordinates (ℓ, b). A WIMP mass of 200 GeV and a WIMP-nucleon cross section of 10^{-46} cm^2 is considered.

Note that the anisotropy of all rates in Eqs. (4), (5), and (6) depends only on the velocity $\mathbf{V}$. In a given frame, which fixes $\mathbf{V}$, one can choose different coordinate systems, yielding different components for $\mathbf{V}$. The coordinates can be also time dependent. In a frame at rest with respect to the Earth and using Galactic coordinates, $\mathbf{V}_{SG}$ is constant and only $\mathbf{V}_{ES}$ rotates with the annual periodicity of the Earth revolution. Since $\mathbf{V}_{ES}$ is an order of magnitude smaller than $\mathbf{V}_{SG}$, the recoil rate points back to a direction that rotates with an opening angle about one tenth of a radiant and annual periodicity around the $\mathbf{V}_{SG}$ direction. In this frame the peaked angular distribution is the main signature of the WIMP signal and allows background reduction. In a coordinate system fixed with a laboratory, the coordinates and, therefore, the apparent direction of $\mathbf{V}$ makes an additional rotation with the periodicity of a sidereal day and an amplitude that depends on the latitude. This specific periodicity is a characteristic signature and provides more background suppression.

Most of our results will be shown for a detector located at the latitude of LNGS in the local coordinate system where the polar axis points in the vertical direction. A reference frame at rest with respect to the Sun is used only to introduce the potentialities of a directional detector, as it has been extensively done in literature [6].

4. Recoil spectrum in galactic coordinates

The general potentialities of a directional detector and, more specifically, the signature in the angular recoil rate of the detector motion through the WIMP wind are best illustrated in Galactic coordinates. In this coordinate system $\hat{x}$ points from the Sun towards the Galactic center, $\hat{y}$ in the direction of the Solar motion and $\hat{z}$ towards the Galactic north pole; therefore, $\mathbf{V} = v_0\hat{y}$. In Fig. 1 it is shown the angular recoil rate of Eq. (4) on a Mollweide equal area projection map in Galactic coordinates. The horizontal axis is the Galactic longitude, 0°$< \ell <$ 360°, and the vertical axis the Galactic latitude, -90°$< b <$ 90°. The recoil rate is clearly anisotropic and points towards coordinates (ℓ=270°, b=0°), opposite to the direction of the Sun motion throughout the Galaxy.

5. Recoil directional signals at LNGS

In this section, the expected rates of WIMP scattering, Eqs. (4-6), is illustrated as a function of $\cos\theta_r$ for different times of an arbitrary chosen day, corresponding to the Summer Solstice

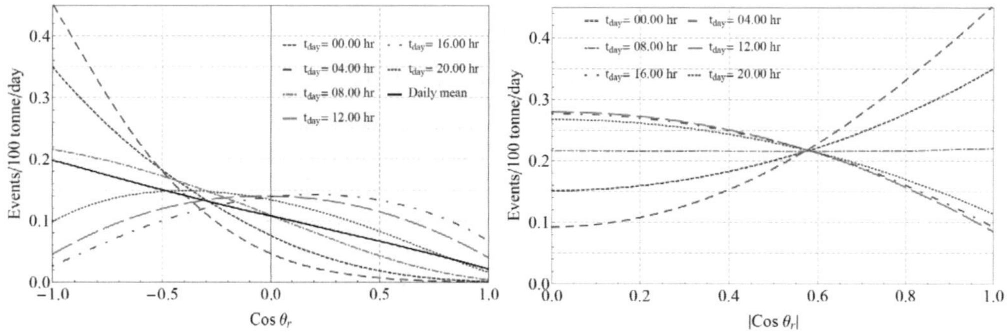

Figure 2. Differential (left) and folded differential (right) recoil rate as a function of the polar angle $\cos\theta_r$ at the latitude of LNGS. The y axis represents events per day in a 100 tonne Argon detector with a 50 keV energy threshold for a cross section $\sigma_n = 10^{-46}$ cm^2. The solid curve is the daily average, while the other six curves show the rate at times of the day four-hour apart.

(SS). Obviously, there is a strict correlation between the expected signal and the position of the Cygnus in the sky as seen at the LNGS latitude since it corresponds to the main direction of the WIMP wind. During most of the day, the Cygnus constellation is over the horizon, reaching the zenith at ~4:00 a.m. and setting around 12 hours later, remaining near the horizon just for few hours. Fig. 2 shows on the left the recoil rate, Eq. (5), as a function of $\cos\theta_r$ for different times of the day and the daily average. There is a strong dependence on the time of the day, since the asymmetry is larger when the Cygnus is high in the sky (00:00 $\leq t_{day}$[hr] $\leq$ 08:00), smaller when it is close to the horizon (12:00 $\leq t_{day}$[hr] $\leq$ 20:00). This implies that it is possible to discriminate the WIMP nature of the events against possible backgrounds. Indeed, even the average rate varies considerably since it changes from ~0.2 events/(100 tonne day) for $\cos\theta_r = -1$, to ~0.02 events/(100 tonne day) for $\cos\theta_r = 1$. In the right panel of Fig. 2 it is shown the "folded" differential recoil rate introduced in Eq. (6), which should be the rate relevant for a CR detector. The angular and time dependences of the rate remain quite strong even without the information on where the track is pointing. At times when the Cygnus is close to the zenith (horizon) the rate is peaked at $|\cos\theta_r| \sim 1$ ($|\cos\theta_r| \sim 0$).

5.1. A realistic CR detector

A realistic CR detector have a finite angular resolution. Here, it is studied how much the angular signature of the signal is reduced assuming to have only a minimal amount of angular information. Events are divided in only two classes. Horizontal events (HOR), defined by $|\cos\theta_r| < 0.5$ or 60°$< \theta_r <$120°, and vertical events (VER=UP+DOWN), defined by $|\cos\theta_r| > 0.5$, as shown schematically in the left panel of Fig. 3. From an experimental point of view, ratios of event rates are useful quantities in order to keep the systematic uncertainties budget under control. Thus, the ratio, R, of HOR and VER events is defined, which should be constant and equal to one for an isotropic signal. On the right panel of Fig. 3 the time dependence of this ratio and of the single components is shown. The dashed lines represent the daily variation of the HOR and VER event rates, and their ratio, R, for four equidistant sidereal days. Use of the sidereal day (s.d.) as a unit is convenient because after a s.d. the Cygnus constellation returns exactly in the same position in the sky and the angle between the WIMP wind and the detector axis has the same value. Apart from a small variation in the velocity of the WIMP wind caused by the Earth revolution around the Sun, a similar behaviour of the HOR and VER components during different sidereal days is expected. Therefore it is possible to make the annual averages of the daily rates in order to accumulate statistics. The

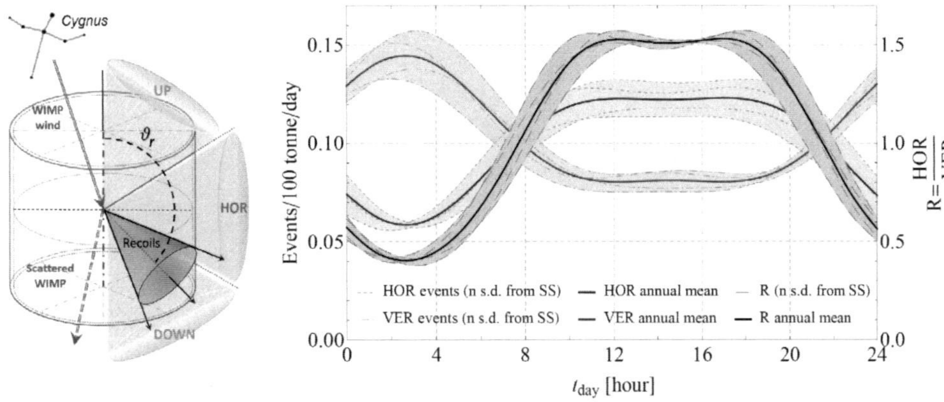

Figure 3. Left: Schematic representation of the recoils distribution between horizontal (HOR $\equiv |\cos\theta_r| < 0.5$) and vertical (VER=UP+DOWN $\equiv |\cos\theta_r| > 0.5$) events in a cylindrical LAr TPC for a particular positions of the Cygnus constellation in the sky. Right: Daily variation of HOR (VER) event rate in blue dashed (orange dotted) lines for different sidereal days of the year. The solid blue (orange) line represents the annual mean of the HOR (VER) events. They refer to the left scale. Ratio of HOR and VER events for different sidereal days (dashed black lines) and the corresponding annual mean (solid black line) refer to the right scale.

results are shown in the right panel of Fig. 3 in which the blue (orange) band represents the overall variation registered in the HOR (VER) component among different sidereal days of the year, while the solid curves represent the averages over different sidereal days. Considering the annual mean, the variation of the single component is of the order of 35% (29%), with respect to the average daily value. For comparison, the seasonal modulation of the events that can be observed under the same assumptions, for the case of a non-directional WIMP detection is only about 8%. Compared to it, the effect of the daily variation in a directional detector, even for this minimal angular resolution, confirms that adding the directional information increases significantly the discovery potential of a WIMP signature. Moreover, the annual mean ratio, R, shows a remarkable deviation from one since it varies during the day by a factor $f \approx 4$.

6. Conclusions

In these proceedings, it is discussed how a directional detector located at LNGS could provide a very clear signature for WIMP signal. This signature remains sizeable even for a minimal angular resolution which should be easily obtained by a detector based on columnar recombination. In particular, the daily modulation shown in Fig. 3 is an order of magnitude larger than the annual modulation obtained ignoring the angular information. Thus, a realistic detector able to use the angular information to reject the background would provide a strong and unambiguous evidence of DM, the latter due to the fact that no known background is correlated with the sidereal period of the DM modulation.

References

[1] Bernabei R et al 2008 *Eur. Phys. J.* C **56** 333-355
[2] Ahlen S et al 2010 *Int. J. Mod. Phys.* A **25** 151
[3] Jaffé G 1913 *Ann. Phys. (Leipzig)* **42** p. 303
[4] Gondolo P 2002 *Phys. Rev.* D **66** 103513
[5] Akerib D S et al 2014 *Phys. Rev. Lett.* **112** 091303
[6] Billard F, Mayet J, Macias-Perez F and Santos D 2010 *Phys. Lett.* B **691** 156-162

Journal of Physics: Conference Series **689** (2016) 012016 doi:10.1088/1742-6596/689/1/012016

Neutron detection in nuclear astrophysics experiments: study of organic liquid scintillators

Giovanni Francesco Ciani[1,2,3]

[1] Università degli Studi di Bari, Piazza Umberto I, 1, 70121 Bari, Italy
[2] INFN - Laboratori Nazionali Del Gran Sasso, Via Acitelli 22, 67100 L'Aquila, Italy
[3] Gran Sasso Science Institute (INFN), Viale Francesco Crispi 7, 67100 L'Aquila, Italy

E-mail: giovanni.ciani@lngs.infn.it

Abstract. In order to study the nuclear reaction $^{13}C(\alpha,n)^{16}O$, crucial for the nucleosynthesis of heavy nuclei (A>58), the LUNA collaboration at *Laboratori Nazionali del Gran Sasso*, is looking for the best neutron detector to use in the set up. One of the possibilities is to use detectors based on cell filled with Organic Liquid Scintillator BC501A. These detectors are sensible to fast neutron, but also to gamma rays. A Pulse Shape Discrimination process using the Zero Crossing method has been performed to select only signals from neutrons. Comparing the neutron spectra after the Pulse Shape Discrimination and the spectrum from a *GEANT4* simulations, the efficiency of the BC501A, in function of the neutron energy and varying the light threshold, has been evaluated.

1. The astrophysical motivation

All the nuclei existing in nature are synthesized, according to their atomic mass A, in several processes, the so-called nucleosynthesis processes, which takes place in different astrophysics scenarios.Heaviest nuclei (A>58) are created by neutron capture. For unstable nuclei, when the time T_n between successive neutron captures is much larger than the beta decay lifetime T_b ($T_n \gg T_b$), the processes involved are called slow process (*s process*), that take place in stars with low mass ($M < 8M_\odot$).
In low-mass Asymptotic Giant Branch (AGB) [2] stars ($1 - 3M_\odot$) the main neutron source is the $^{13}C(\alpha,n)^{16}O$ reaction, the neutrons produced are fuel of the *s-process* producing about half of the stable isotopes beyond iron in the Universe.
The temperatures involved in these processes are 90 - 100 MK, which roughly correspond to Gamow energies between 180 and 200 keV. At present, the cross section within the Gamow peak is uncertain by almost one order of magnitude, having a large impact on stellar models. The cross section σ is strongly dependent from the energy in the center of mass E and can be factorized as:

$$\sigma(E) = \frac{1}{E} e^{-2\pi\eta} S(E) \quad (1)$$

where the exponential represents the probability to overcome the Coulombian barrier of a certain reaction, $\eta = \frac{Z_1 Z_2 e^2}{4\pi\epsilon_0 \hbar v}$ is the Sommerfeld parameter, where Z_1e and Z_2e are the charges of the two nuclei involved in the reaction and v is their relative velocity and finally $S(E)$ is the so called astrophysical S factor, that contains all the nuclear effects of the reaction and varies less rapidly

with the energy E than the cross section [3].
The state of the art, shown in Fig. 1, presents measurements of the astrophysical factor as a function of the energy of the direct reaction at energy greater than the Gamow Window [4, 5] and extrapolations performed using indirect techniques [6–8] that differ by factor 3-5. For this reason, a renewed direct study of the reaction $^{13}C(\alpha,n)^{16}O$ at energies below $E_{cm} = 300keV$ is demanded. At these low energies, the cross section is so low that the reaction rate should be really poor and, in surface laboratories, the events of interest should be hidden by the background coming from cosmic rays. So, it is needed to go underground in order to have a natural shielding. This is the reason why this reaction will be studied at National Laboratory of Gran Sasso (LNGS) in the scientific framework of the experiment *Laboratory for Underground Nuclear Astrophysics* (LUNA) both at the accelerator currently working (that has a maximum electrostatic potential of 400 kV), and at the future facility $LUNA\ MV$[9].

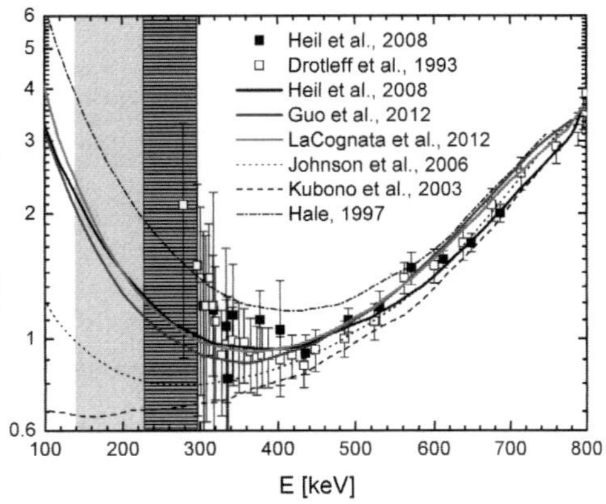

Figure 1: State of the art: the S(E)-factor data [4–8] of the $^{13}C(\alpha,n)^{16}O$ reaction.

2. The neutron detection and the Pulse Shape Discrimination

The $^{13}C(\alpha,n)^{16}O$ reaction emits fast neutrons with energy between 2-3 MeV: their detection plays a crucial role in the precise measurement proposed by LUNA collaboration, where the uncertainties foreseen are between 5% and 12% for data points in the energy range between 230 keV and 400 keV.
In this contest, a neutron detector based on Organic Liquid Scintillators BC501A has been tested. The structure is shown in Fig. 2 [10].

This detector is sensible to fast neutrons (detected by mean of their elastic scattering with protons in the cell), but also to gamma rays, that mainly interact with electrons in the Compton Scattering process. For this reason, a *Pulse Shape Discrimination* (PSD) process has been performed to select only signals due to neutrons, using the Zero Crossing method [11, 12]: the passage of different type of particle inside the scintillators causes a different signal of scintillation yield. Heavier is the particle, greater is the slow component of the anode input of the photomultiplier and longer is the time difference T between the zero levels of the bipolar signal shaped by a CR-RC-CR circuit, as shown in Fig. 3 [13]. After a proper calibration of the set up in energy and in time, the PSD was performed using an AmBe source: the results are shown in Fig. 4, where the neutrons are located in the upper cluster, while the gammas in the lower one. Note that the energy scale is labelled as MeV_{ee}, MeV electron equivalent, meaning that the energy scale is calibrated on γ particles, neglecting the neutron quenching factor. The two clusters are discriminated using a threshold parametrized with a quadratic function, whose

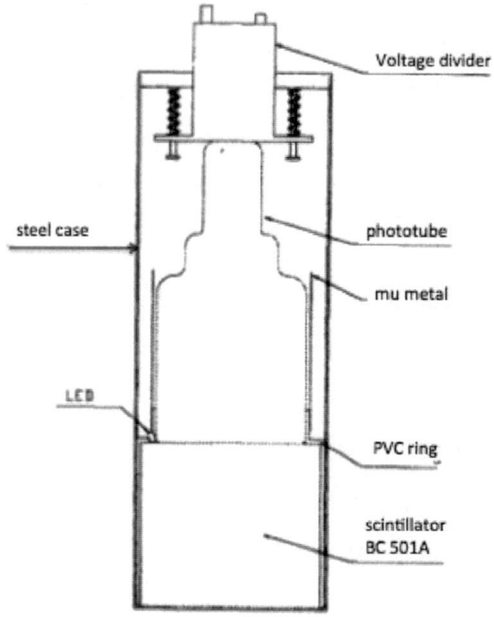

Figure 2: Scheme of the neutron detector tested [10].

trend favors the best way to disentangle, in an offline analysis, the signals due to neutrons (n) and the gammas (γ) from the background.

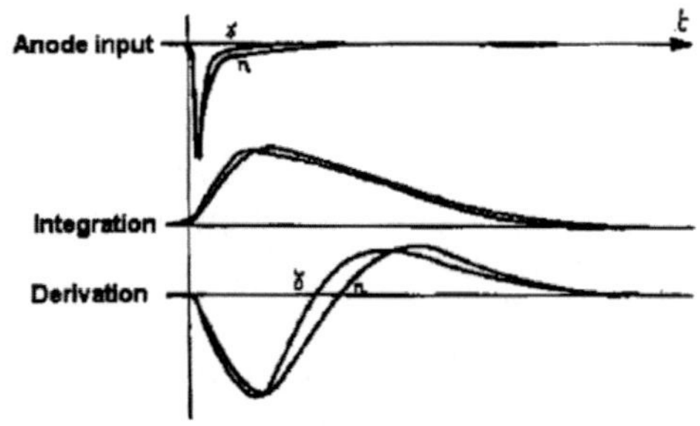

Figure 3: Both a gamma ray and a neutron in a scintillator produce a negative unipolar signal: these are shaped by a RC-CR-RC in a bipolar signal. The difference between the two zero points is grater for the neutron[12].

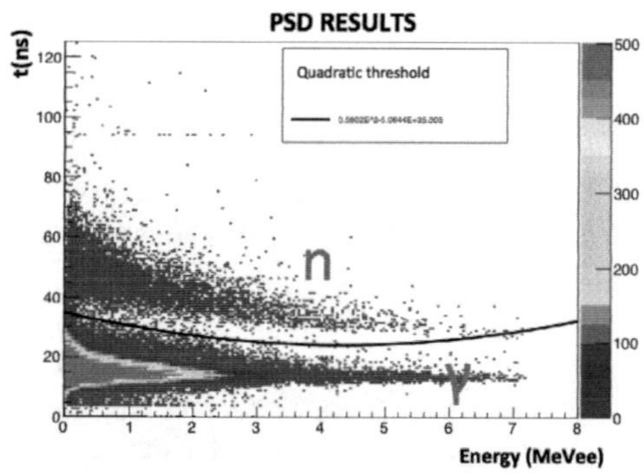

Figure 4: T vs Energy of the particle: two different clusters, regarding neutrons (up) and gammas (down), are evident.

3. Simulation study and Detection efficiency

In parallel to the experimental PSD, a dedicated GEANT4 [14] Monte Carlo code of the system AmBe source plus BC501A scintillator has been carried out. The experimental results and the simulation have been compared as shown in Fig. 5: the fitting between the two spectra permits to conclude that the discrimination is performed with good results.
It is also possible to estimate the neutron efficiency of the BC501A detector in function of the neutron energy and varying the light threshold, as shown in Fig. 6: the simulation yielded an efficiency of 50% at 2 MeV, the energy region of interest for the reaction $^{13}C(\alpha,n)^{16}O$.

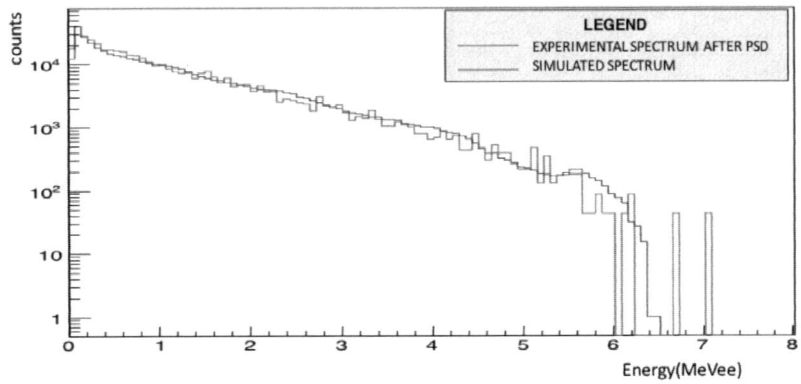

Figure 5: Comparison between the neutron experimentally discriminated after the PSD process and the GEANT4 simulation.

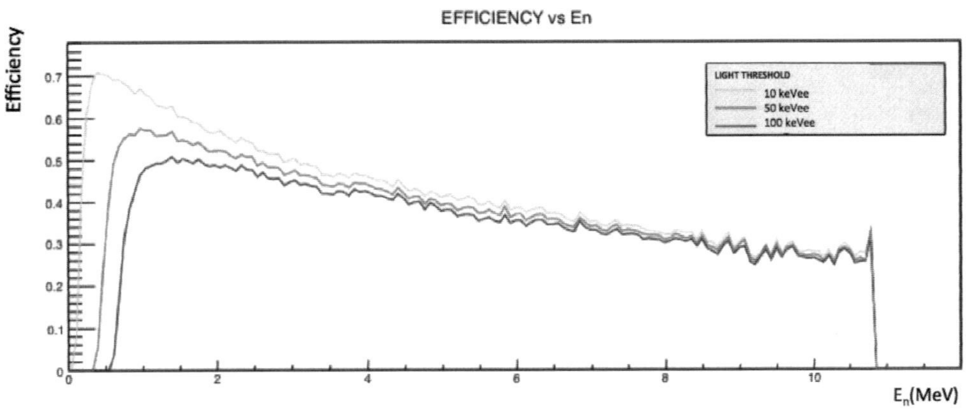

Figure 6: Neutron detection efficiency E_n in function of incident neutron energy and varying the threshold.

The results are encouraging even if the work is still in progress. Several geometries and possible compact array should be considered to answer the experimental requests as high efficiency in the MeV range according to a flat distribution and low intrinsic background (i.e. selected materials).

Journal of Physics: Conference Series **689** (2016) 012016 doi:10.1088/1742-6596/689/1/012016

References

[1] Burbidge E M, Burbidge G R, Fowler W A and Hoyle F 1957 *Rev. Mod. Phys.* **29**(4) 547–650
[2] Rolfs C and Rodney W 1988 *Cauldrons in the Cosmos: Nuclear Astrophysics* (University of Chicago Press)
[3] Drotleff H *et al.* 1993 *Astrophys. J.* **414** 735
[4] Heil M *et al.* 2008 *Phys. Rev. C* **78**(2) 025803
[5] Johnson E D *et al.* 2006 *Phys. Rev. Lett.* **97**(19) 192701
[6] Pellegriti M G *et al.* 2008 *Phys. Rev. C* **77**(4) 042801
[7] La Cognata M *et al.* 2012 *Phys. Rev. Lett.* **109**(23) 232701
[8] Guglielmetti A 2014 *Phys. of the Dark Univ.* **4** 10 – 13 ISSN 2212-6864
[9] Colonna N *et al.* 1996 *Nucl. Instr. Meth. Phys. Res. A* **381** 472 – 480 ISSN 0168-9002
[10] Knoll G F 2000 *Radiation Detection and Measurement* (Wiley) ISBN 9780471073383
[11] Sperr P, Spieler H, Maier M and Evers D *Datasheet Pulse shape discriminator Model 2160* CANBERRA
[12] Winyard R, Lutkin J and McBeth G 1971 *Nucl. Instr. Meth.* **95** 141 – 153 ISSN 0029-554X
[13] Agostinelli S, Allison J, Amako K, Apostolakis J *et al.* 2003 *Nucl. Instr. Meth. Phys. Res. A* **506** 250 – 303 ISSN 0168-9002

Development and tests of a new prototype detector for the XAFS beamline at Elettra Synchrotron in Trieste

S Fabiani[1], M Ahangarianabhari[2],G Baldazzi[3], P Bellutti[4], G Bertuccio[2,5], M Bruschi[6], J Bufon[7,1,8], S Carrato[8], A Castoldi[5,9], G Cautero[7,1], S Ciano[1],A Cicuttin[10] , M L Crespo[10], M Dos Santos[10] , M Gandola[2], G Giacomini[4], D Giuressi[7], C Guazzoni[5,9], R H Menk[7,1],J Niemela[10], L Olivi[7,1], A Picciotto[4], C Piemonte[4], I Rashevskaya[11], A Rachevski[1], L P Rignanese[3], A Sbrizzi[6], S Schillani[7], A Vacchi[1,12], V Villaverde Garcia[10], G Zampa[1], N Zampa[1], N Zorzi[4]

[1] Istituto Nazionale di Fisica Nucleare, Section of Trieste, Padriciano 99, I-34149, Trieste, Italy
[2] Politecnico di Milano, Via Anzani 42, I-22100, Como, Italy
[3] Universitá di Bologna, Via Zamboni 33, I-40126, Bologna, Italy
[4] Fondazione Bruno Kessler - FBK, via Sommarive 18, I-38123, Trento, Italy
[5] INFN Milano, Via Celoria 16, I-20133, Milano, Italy
[6] INFN Sezione di Bologna, Via Irnerio 46, I-40126, Bologna, Italy
[7] Elettra-Sincrotrone Trieste S.C.p.A., SS14, I-34012, Trieste, Italy
[8] Universitá di Trieste, Piazzale Europa 1, I-34128, Trieste, Italy
[9] Politecnico di Milano, Piazza Leonardo da Vinci 32, I-20133, Milano, Italy
[10] International Centre for Theoretical Physics - MLAB, Via Beirut 31,I-34014, Trieste, Italy
[11] TIFPA-INFN, via Sommarive 14, I-38123, Trento, Italy
[12] Universitá di Udine, Via delle Scienze 206, I-33100, Udine, Italy

E-mail: `sergio.fabiani@ts.infn.it`

Abstract. The XAFS beamline at Elettra Synchrotron in Trieste combines X-ray absorption spectroscopy and X-ray diffraction to provide chemically specific structural information of materials. It operates in the energy range 2.4-27 keV by using a silicon double reflection Bragg monochromator. The fluorescence measurement is performed in place of the absorption spectroscopy when the sample transparency is too low for transmission measurements or the element to study is too diluted in the sample. We report on the development and on the preliminary tests of a new prototype detector based on Silicon Drift Detectors technology and the SIRIO ultra low noise front-end ASIC. The new system will be able to reduce drastically the time needed to perform fluorescence measurements, while keeping a short dead time and maintaining an adequate energy resolution to perform spectroscopy. The custom-made silicon sensor and the electronics are designed specifically for the beamline requirements.

1. Introduction

The XAFS (X-ray absorption fine structure) beamline at Elettra Synchrotron in Trieste combines the capability of X-ray absorption spectroscopy (XAS) with X-ray diffraction (XRD) to provide chemically specific structural information. Both techniques allow to fully characterize material

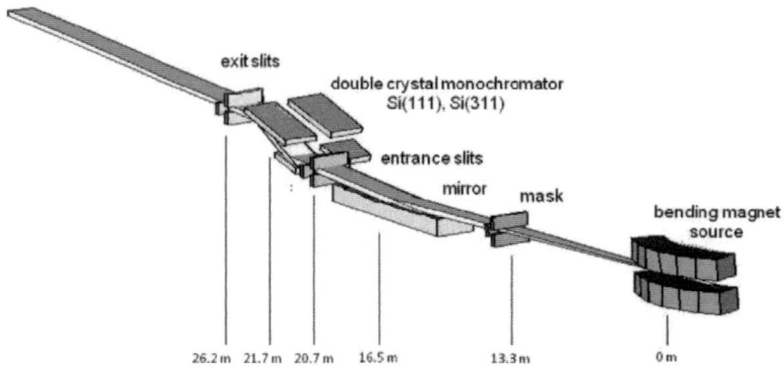

Figure 1: Scheme of the XAFS beamline. Figure from [1]

structure. The XAFS spectrometer [1] is located along the tangential fan of a bending magnet as shown in the sketch of Fig. 1. The monochromator (a double flat crystal, double cam Kohzu apparatus) allows the XAFS beamline to operate in the energy range 2.4–27 keV by using an interchangeable pairs of Si(111) and Si(311) crystals on which X-rays undergo two successive Bragg reflections. Four tungsten alloy slits placed after the monochromator are used to define the beam section impinging on the sample.

The beamline is designed for spectroscopic analysis aimed at studying the behaviour of the absorption coefficient as a function of energy in the XAS region (absorption spectroscopy). The energy scan is performed typically in an energy range of 1–2 keV around the energy of interest by operating on the Si double crystal monochromator, so to sample between 1000 and 2000 energy points.

However, the fluorescence analysis is performed in place of the absorption spectroscopy when the sample is optically thick and does not allow for transmission measurements. With the present set-up about 5–10 s of integration are needed to have enough statistics for each energy value. Therefore, up to 2×10^4 s (about 6 hours) are needed to sample 2000 energy points.

A sample typically comprises a matrix of various elements in addition to the one of interest. All of them could be excited via fluorescence. Moreover, radiation scattering can occur on the set-up spreading photon energy across the energy spectrum. Due to these effects sometimes only a very small fraction of the emitted radiation from the specimen comes from the elements of interest. Therefore, the detector must be able to sustain a high count rate with an adequate energy resolution to perform the measurements. Currently, a KETEK GmbH AXAS-M Silicon Drift Detector (SDD) is employed for the fluorescence measurements. It has a single 100 mm^2 SDD cell (80 mm^2 of effective area with collimation) with an FWHM (Full Width Half Maximum) energy resolution of $\sim$170 eV for the Mn Kα line at 5.89 keV for a peaking time of 1$\sim$ μs at -70°C. This set-up cannot sustain a high count rate. This is mainly due to the dead time caused by the monolithic geometry of the Si sensor.

This work reports on the study carried out to design a new fluorescence detector, based on the Silicon Drift Detector technology, able to maintain a short dead time and an adequate energy resolution to perform spectroscopy. This goal will be achieved by manufacturing a new custom-made detector designed specifically for the beamline requirements.

2. The detector

The new detector will have a larger area and will be segmented to maintain the pile-up at low level. A sensor will comprise small cells each one constituting an independent read-out channel.

Journal of Physics: Conference Series **689** (2016) 012017 doi:10.1088/1742-6596/689/1/012017

Small cells will have a low leakage current, of the order of 10 pA at ambient temperature, allowing to perform measurements with a moderate cooling. Therefore, the cooling system will not be a critical issue even for the large area detector. Its final version will be manufactured to work both in air and in vacuum.

Figure 2: SDD linear array comprising 8 squared cells 3×3 mm^2. Anode (top) and entrance window (bottom) sides.

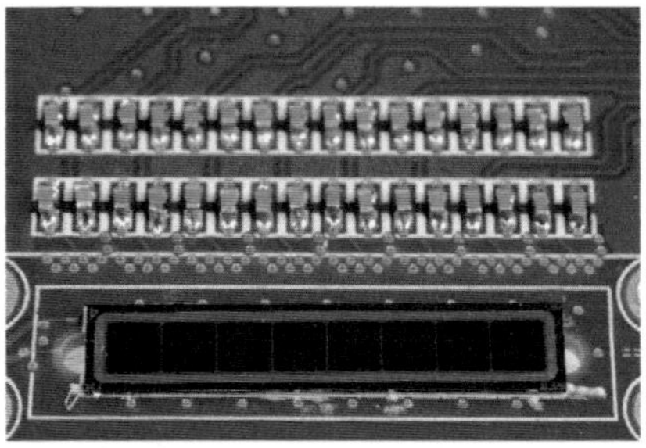

Figure 3: The SDD front side mounted on the front-end electronics board.

Figure 4: Back-end electronics with SDD back side bonded to the preamplifiers.

Figure 5: Front end electronics board with the plastic box enclosing the SDD.

In this work we report on the design activity and tests of the prototype detector given by a single SDD sensor (representative of the final version) coupled to prototypal front-end and back-end electronics boards. We developed a simulation software based on the GEANT4 toolkit [2] to optimize the detector design. The simulator reproduces the beamline source energy spectrum and the beam shape out of the last set of four tungsten alloy slits before the sample. To use the detector at the highest count rate conditions, that is measuring fluorescence from thick foils of pure elements, we estimated that each detector channel must be able to sustain a maximum data rate of the order of 10^5 counts s^{-1}. A single SDD sensor module (see Fig. 2) is designed as an array of 8 squared cells 3×3 mm^2. The prototype detector comprises a single SDD array (see Fig. 3), while the final version will be made of 8×8 matrix for a total collecting area of about 576 mm^2, seven times larger than the current detector.

The SDD sensors of the new detector are obtained from high resistivity n-type silicon wafers 450 μm thick. They are designed by INFN-Ts and manifactured by FBK. Their layout is such that the energy of an absorbed X-ray photon is converted into electron-hole pairs and electrons are drifted towards the anodes (small n$^+$ pad) on the back side at the center of each cell. The drift cathodes on the back side are arranged as a decreasingly negative biased p$^+$ rings. At the border of each cell the outermost drift cathode is kept at the bias voltage and voltage dividers are integrated separately for each cell to generate potential drops down to the innermost one.

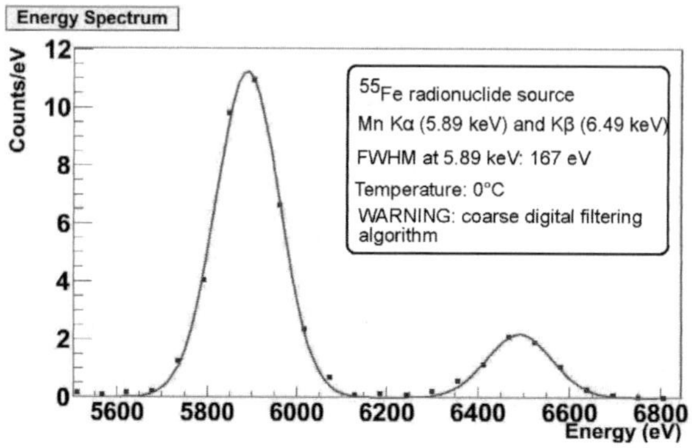

Figure 6: Energy spectrum of an ^{55}Fe source at 0°C measured in the climatic chamber.

The entrance window is biased separately with respect to the detector back side allowing for an effective charge collection. Both on the back and the front side, outside the sensitive area of the whole cell array, there are guard cathodes that scale down the bias voltage to ground. The readout anodes are wire bonded to ultra-low noise SIRIO charge sensitive preamplifiers on the front-end board (see Fig. 4). The SIRIO preamplifier is designed for SDDs readout in CMOS technology. It has intrinsic minimum noise levels of 1.2 and 0.9 electrons r.m.s. at +22°C and -30°C, which correspond to line's FWHM of about 10 eV and 8 eV, respectively, referring to silicon detectors [3]. This performance represents the current state of the art in low-noise front-end electronics for SDDs and the final energy resolution is practically determined by the detector anode dark current and by the parasitic capacitance of the preamplifier-detector connection. An intrinsic noise as low as 8.6 electrons r.m.s., corresponding to 74 eV FWHM on the pulser line, has been demonstrated with a 13 mm^2 SDD-SIRIO system operating at room temperature [4]. The used ASIC is composed by SIRIO and an output amplifier able to drive a 50 Ω coaxial cable up to 2 m length with a total power consumption of about 10 mW.

The output of the preamplifiers are low-pass filtered to provide anti-aliasing for the following 12 bit 8-channel ADC, capable of encoding at 40 Msps. The digital data are subsequently treated with a set of suitable digital filters and analysed by an FPGA (Field-Programmable Gate Array), which also handles the near-saturation reset of the preamplifiers. The reset of all the preamplifiers is simultaneous and it is driven by the first one reaching the near-saturation threshold. The acquired data are transmitted over a TCP / IP connection to a PC running a dedicated LabVIEW software. The back-end FPGA encodes also a shift register hosted on the front-end electronics that allows to disable separately each one of the eight channels by holding the corresponding preamplifier in a reset state. Since the sensor is sensitive to light, a small plastic box was designed to enclose and keeping it in a dark environment. The X-ray entrance window was realized stacking two foils 25 μm thick of alumined Mylar (see Fig. 5).

3. Tests in laboratory and measurements at the beamline

A series of characterization tests were performed on the prototype detector in laboratory at ambient temperature and in climatic chamber down to 0°C. The energy resolution was measured by using a coarse off-line algorithm (not yet optimized) for the data analysis. In Fig. 6 the energy spectrum of an ^{55}Fe source at 0°C measured in the climatic chamber is shown. The energy resolution of the Mn Kα line at 5.89 keV is 167 eV.

A beam test with the prototype detector was carried out at the XAFS beamline of Elettra in the period 16-18 September 2015. The measurement set-up is shown in Fig. 7. A foil of Mn (Kα

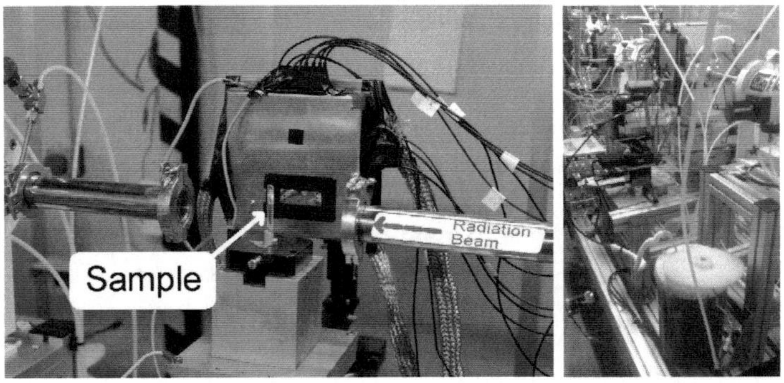

Figure 7: On the left: Measurement set-up of the beam test at the XAFS beamline. A foil of a pure element (Mn or Zr) is excited by the radiation beam and emits fluorescence photons. On the right: the dewar filled with liquid nitrogen used to refresh the nitrogen flux that allows the moderate cooling.

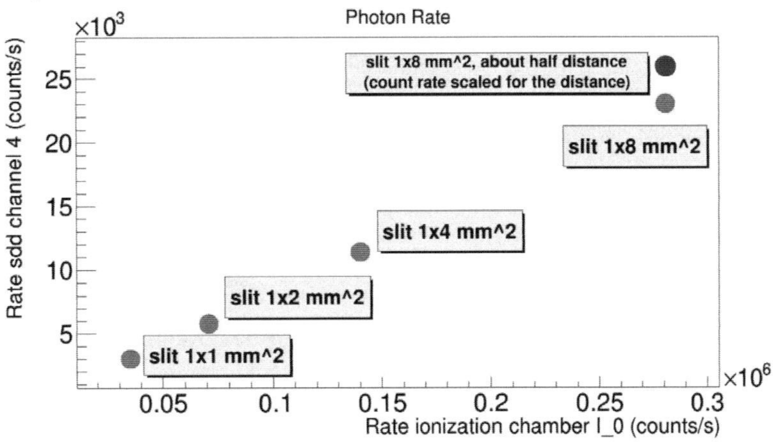

Figure 8: Measured count rate on a single SDD cell as a function of the count rate of the first ionization chamber of the XAFS beamline. The highest count rate (blue point) was scaled by a factor $1/4^{th}$ (see Fig. 9) to be compared to the other measurements (red points) taken at a larger SDD/sample distance.

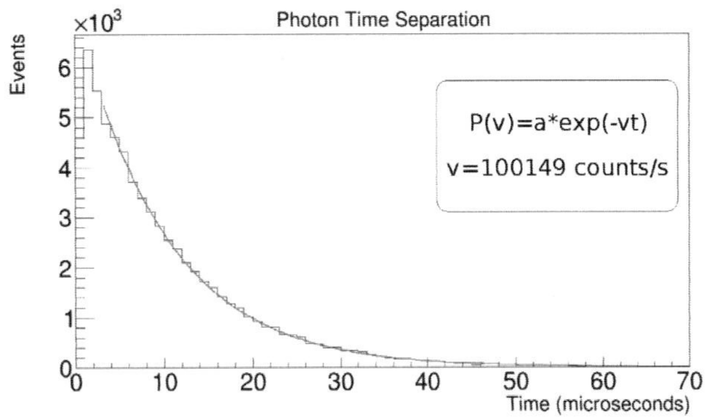

Figure 9: Exponential fit of the Poisson distribution of photon arrival time with a slit opening of 1×8 mm^2 and 5.5 cm of distance between the sample and the SDD. A count rate of about 10^5 counts/s was reached without any significant limitation due to pile-up.

at 5.89 keV and Kβ 6.49 keV) and then Zr (Kα at 15.75 keV and Kβ 17.67 keV) were placed in front of the detector a 45° with respect to the beam axis. The front-end board hosting the SDD was inside an aluminum shielding box which entrance window was covered with a 25 μm thick alumined Mylar foil to stop stray visible light entering in the box. The three alumined Mylar foils (two on the entrance window of the plastic box on the front-end electronics board and the third one on the aluminum box), have a total thickness of 75μm for a transparency of 84.5% at 6 keV. The prototype detector was operated at nearly ambient temperature by fluxing in the plastic box a moderately cooled nitrogen flux. This system allowed to remove the heat produced by electronics components and preamplifiers enclosed in the plastic box.

We tested the detector functionality at different photon fluxes. The photon flux on the sample, and thus the count rate of the detector, depend on the opening of the last set of slits that was increased by steps. Moreover, the count rate was further increased by halving the distance between the sample and the SDD from about 10.8 cm to about 5.5 cm. Using this set-up, we reproduced a very high count rate condition, far away from the working capabilities of the current detector. In Fig. 8 we report the measured count rate with the Mn sample on a single SDD cell as a function of the count rate of the first ionization chamber that is crossed by radiation before impinging on the sample. The red points were obtained placing the detector at 10.8 cm from the sample. The blue one correspond to the count rate measured by placing the detector close to the sample (5.5 cm) and corrected to take into account the different detector to sample distances. Therefore, this count rate value of 100149 counts/s (see Fig. 9) was scaled by the ratio of the squares of the two distances ($1/4^{th}$). Since the scaled value is not smaller than the value obtained at 10.8 cm, we conclude that this measurement is not limited by pile-up.

The beam test was successful and detailed results will be published in a forthcoming article.

4. Conclusions

We developed and tested a prototype detector for the XAFS beamline at Elettra Synchrotron in Trieste based on the SDD technology. The prototype demonstrated the importance of designing a modular detector to reach a larger collecting area while maintaining low pile-up and short dead time. The prototype detector comprises a single SDD array of eight cells, each one coupled to a SIRIO ultra low noise preamplifier to form an independent channel. The detector final version will be an 8×8 matrix for a total collecting area of about 576 mm^2, seven times larger than the detector currently employed at the beamline. During preliminary tests, we measured the performance of the prototype detector in laboratory and obtained an energy resolution of 167 eV at 5.89 keV at 0°C. We also performed a test beam on the prototype detector at the XAFS beamline of Elettra and verified its performance for high count rate.

Acknowledgments

We acknowledge EUROFEL-MIUR and INFN R&D ReDSoX projects for founding this activity.

References

[1] Di Cicco A, Aquilanti G, Minicucci M, Principi E, Novello N, Cognigni A and Olivi L 2009 *Journal of Physics Conference Series* **190** 012043

[2] Geant4 Collaboration 2010 Geant4: A Simulation Toolkit for the Passage of Particles through Matter astrophysics Source Code Library (*Preprint* 1010.079)

[3] Bertuccio G, Macera D, Graziani C and Ahangarianabhari 2014 *Proceedings of the IEEE Nuclear Science Symposium Seattle WA - USA, 8-15 Nov.*

[4] Bertuccio G, Ahangarianabhari M, Graziani C, Macera D, Shi Y, Rachevski A, Rashevskaya I, Vacchi A, Zampa G, Zampa N, Bellutti P, Giacomini G, Picciotto A and Piemonte C 2015 *Journal of Instrumentation* **10** P01002

Journal of Physics: Conference Series **689** (2016) 012018 doi:10.1088/1742-6596/689/1/012018

The FAMU experiment: muonic atoms to probe the proton structure

D Guffanti[1] on behalf of the FAMU collaboration

[1]INFN Trieste, Area di Ricerca, Padriciano 99, I-34149, Trieste (TS) and Gran Sasso Science Insitute - INFN, viale F. Crispi 7, I-67100, L'Aquila (AQ)

E-mail: daniele.guffanti@gssi.infn.it

Abstract. The goal of the FAMU experiment is the measurement of the proton Zemach radius using muonic hydrogen, a subject that has raised much interest in recent years due to its implications in the so-called proton radius puzzle. In order to extract the Zemach radius, the FAMU collaboration aims at measuring the hyperfine splitting of the μp ground state, since the effect of the proton finite size affects the HF transition energy. The proposed experimental method requires a detection system which is suited for time resolved X-ray spectroscopy: in this contribution the results of the first measurements performed at the RIKEN-RAL muon facility in order to verify the fitness of the detection system in the pulsed intense muon beam are presented. The characteristic X-rays from atomic transitions in muonic atoms formed in different targets have been detected using a HPGe detector and five scintillating counters based on $LaBr_3(Ce)$ crystals, whose output has been recorded for 5 µs using a 500 MHz digitizer to measure both the energy and the time spectrum of the detected events. With a detailed pulse analysis considering pile-up events, both the expected characteristic X-rays and lifetimes of various elements were measured, paving the way for future measurements to be carried out in early 2016.

1. The strange case of the proton radius

The proton is one of the most common particles in the universe and one of the building blocks of ordinary matter; it is also one of the first particles discovered in 1919 by Ernest Rutherford. Since then, the proton has been widely studied but it still manage to astonish physicists.

Until 2010, there were only two ways to dig with high precision into the electromagnetic structure of the proton in order to get information on its size: the first approach is the study of elastic scattering of electrons and positrons on hydrogen nuclei and accurate measurements of the Lamb shift in the hydrogen atom spectrum. Indeed, the Lamb shift groups all the contributions in the energy level that aren't taken into account in the "standard" quantum treatment, such as QED and recoil corrections and the effects due to the nuclear structure, like its finite size. The level of accuracy needed to measure the effect of the nuclear size (namely the root mean square of the charge distribution) was reached in the 1990s, providing a value of the proton charge radius totally consistent with the ones obtained with scattering experiments[1]. In order to increase the accuracy of the this value, a measurement of the Lamb shift in the $2S$–$2P$ transition of *muonic* hydrogen (μp, *i. e.* the bound state formed by a proton and a negative muon) was proposed; indeed the muon, being about 200 times heavier than the electron is also 200 times closer to

[1] The most recent value computed by the Committee on Data for Science and Technology (CODATA) as a weighted average between various experiment is $r_{\rm p} = 0.8751(61)$ fm [1].

the proton, and thus much more sensitive to its structure. The results of this measurement were published in 2010 and provided a value of the proton charge radius $r_p = 0.841\,84(67)$ fm that was about an order of magnitude more accurate than previous measurements but totally inconsistent [2].

This result was confirmed by other measurements published in 2013 [3], but nobody has been able to explain the discrepancy between "electronic" and "muonic" measurements and all the hypotheses are still on the table, ranging from experimental accident to effects of unconsidered contributions in the calculations or hints for new physics [4].

2. Concept of the FAMU experiment

To help shedding some light on this puzzle, the FAMU project aims at studying the proton structure using muonic atoms, but instead of the charge radius, the goal of the experiment is the measurement of the Zemach radius[2] of the proton R_p which can be extracted performing a precise measurement of the hyperfine splitting of the μp ground state [6, 7]. This quantity has been already measured using ordinary hydrogen, and a comparison with the value extracted from muonic hydrogen may either reinforce or delimit the proton radius puzzle: a substantial agreement with the value of R_p obtained with ordinary hydrogen could suggest that the explanation of the puzzle may lie in unconsidered methodology uncertainties, while a big discrepancy would give good reasons to look for new physics beyond the Standard Model.

The experimental method for this measurement, proposed in [8], combines elementary particles with laser spectroscopy techniques and its schematic representation is shown in Fig. 1. Muons slowed down and stopped in a hydrogen gas target form muonic hydrogen atoms: $\sim 75\%$ in the triplet state ($F = 1$) and the remaining part in the singlet configuration ($F = 0$). Collisions between muonic hydrogen atoms and H_2 molecules quickly de-excite the $\mu p(1S)^{F=1}$ atoms to the singlet state, leaving thermalized muonic hydrogen atoms in the $(1S)^{F=0}$ state. At this point, a laser tuned on the HFS resonance is sent inside the target, inducing a series of singlet-to-triplet transitions. Muonic hydrogen atoms in the $(1S)^{F=1}$ state are once again de-excited back to the singlet state in collision with H_2 molecules and the transition energy is converted into additional kinetic energy of the μp–H_2 system. In this way the μp atom gains about 2/3 of the hyperfine transition energy (≈ 120 meV).

To detect these "kicked" μps, Bakalov et al. [9] proposed to exploit the muon transfer from muonic hydrogen to another higher-Z gas. Indeed, although theory predicts the muon-transfer rate at low energies λ_{pZ} to be energy independent, there are few gases in which it is proved that this is not the case. The first gas that was demonstrated to show such a particular behaviour was oxygen [10, 11], that exhibits a sort of peak in the muon transfer rate $\lambda_{pZ}^{\mathrm{epith}}$ at the epithermal energy (~ 100 eV). More recent theoretical and experimental studies suggested that also argon and neon could exhibit similar properties [12, 13]. Thus, adding small quantities of one of these gases to hydrogen, one can obtain the number of accelerated μps from the number of muon-transfer events measuring the characteristic X-rays of the added gas that are emitted few ns after the formation of the muonic atom.

Performing a scan over the laser frequency near the HFS transition one and counting for each frequency the number of the muon transfer events, one can then obtain a resonance plot and thus the value of the hyperfine splitting of the $1S$ state.

3. The 2014 beam test

The proposed experimental strategy requires a multiple efforts on many different items, not last being the development of a detection system suited for high-rate X-ray spectroscopy; indeed,

[2] The *Zemach radius* was defined by A. C. Zemach in 1956 as the convolution of the charge and magnetic moment density [5].

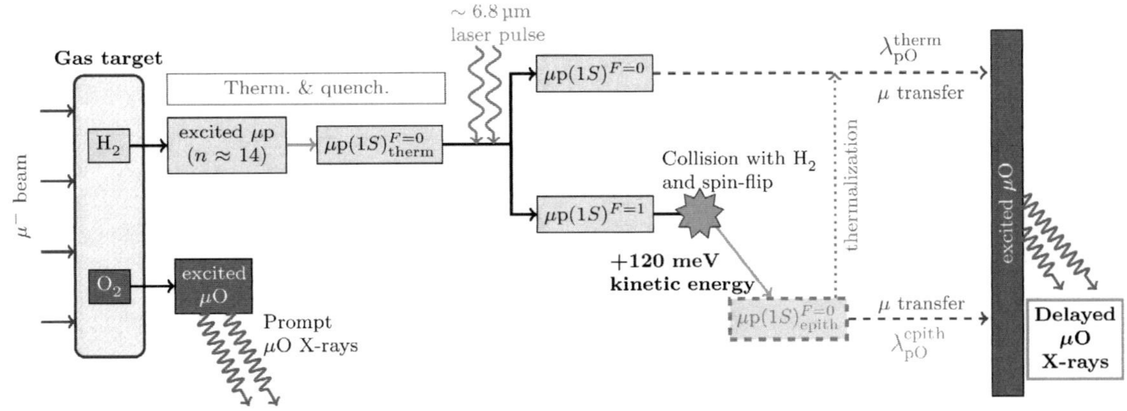

Figure 1: Schematic representation of the FAMU experimental method.

the detection apparatus is not important just in sight of the final experiment, but is also crucial for a series of measurements that are needed in order to validate the method described in the previous section.

3.1. The experimental set-up

A first test of the apparatus was performed in the summer of 2014 at the Rutherford–Appleton Laboratories (UK) where the RIKEN–RAL muon complex [14] is located, the only facility in the world able to provide the pulsed muon beam which is needed for the experiment. An intense muon beam (up to 10^5 muons/(s$\cdot$ cm^2)) is generated by pion decays produced by 800 MeV protons colliding on a carbon target once they have been accelerated by the ISIS synchrotron. The muon beam is then delivered to four experimental ports, and its shape reflects the behaviour of the ISIS proton beam showing a 50 Hz double-pulse structure. Each pulse is about 70 ns long and the time between the two pulses is 320 ns.

Four different targets were exposed to the muon beam: a pure graphite block and three gas mixture (pure H_2, H_2+2%Ar, H_2+4%CO_2, where the percentages are to be intended by weight) contained into an aluminium vessel.

Muonic atoms characteristic X-rays were detected using two different kind of detectors: the heart of the detection system consisted in five scintillating counters based on $LaBr_3$(Ce) crystals, while two HPGe detectors were used in order to have a benchmark spectrum, since they are too slow for the high rate environment. Lanthanum bromide crystals have been chosen for their outstanding performance in terms of energy resolution (2.6% at 662 keV) and decay time ($\tau = 16$ ns), which make them the best inorganic scintillators currently available. Four of the five cylindrical crystals, having diameter of 0.5″ and height 0.5″, were placed in a $80 \times 80 \times 200$ mm^3 iron box coated on its side with a 2 mm thick lead sheet, forming a 2×2 matrix crystals readout by four Hamamatsu R11265-200 PMTs. The other counter was a commercially available Brillance 380 by Saint-Gobain Crystals with embedded PMT reading out a diameter 1″ × 1″ crystal.

To avoid losing useful information, the whole detector output was recorded for 5 µs after the trigger provided by the beam line, using a 500 MHz 14-bit digitizer (CAEN DT5730). The waveforms were then processed off-line with a dedicated software; this choice was also motivated by the specific requirements of the measurement: on one side, not only the energy spectrum of the detected X-rays is interesting but also their time distribution, that can be easily obtained from the digitizer output once a proper analysis method is implemented [15]. On the other hand, from the analysis of the full waveform of the signal one can discriminate single-pulse (Fig. 2a) from pile-up events (Fig. 2b), and may recover information from pile-up events that otherwise

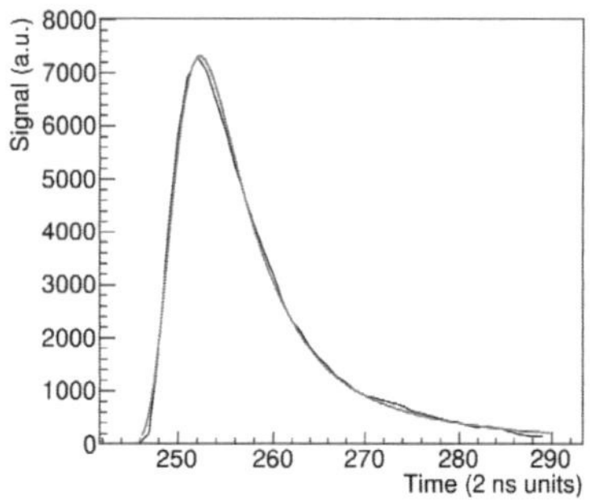

(a) Single pulse

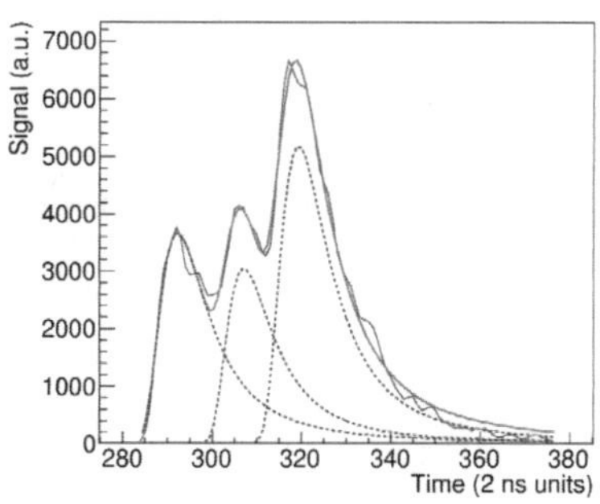

(b) Multiple pulses

Figure 2: Examples of single- **(a)** and multi- **(b)** pulse events as obtained by one of the $LaBr_3$(Ce) counters and processed off-line.

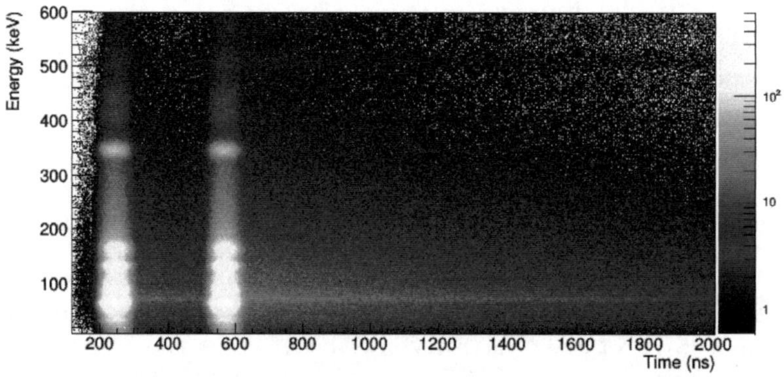

Figure 3: Example of time evolution of the X-ray spectrum for the H_2–CO_2 target. The double pulse structure of the muon beam is evident.

would be lost. Indeed, performing a fit of the signal output, pile-up events were identified and disentangled, recovering more of the 25% of the full data set.

3.2. Data Analysis

Once every pulse induced in the $LaBr_3$(Ce) detectors has been reconstructed, it is easy to extract its starting time and its integral, so to build the time evolution of the X-ray spectrum (Fig. 3).

Once reconstructed the energy spectrum it was possible to look for muonic atoms characteristic X-rays; since muonic atoms are usually formed in highly excited states and quickly de-excite to the ground state emitting photons in the keV range.

The characteristics of the detection system did not allow to reconstruct the energy spectrum below 60 keV, so μp characteristic X-rays (~ 2 keV) are well below the detectability threshold of our system. However, the X-rays emitted by the atomic transitions in heavier atoms and listed in Table 1 were detected.

Because of its peculiar behaviour in the muon transfer process, the detection of oxygen X-rays was one of the goal of the beam test. In the figure a close-up of the region of interest of the spectrum recorded with the four smaller $LaBr_3$(Ce) counters is shown: the K_α line (133.5 keV) is well resolved and also the other K transition lines are present, though not resolved because of the limited energy resolution, as shown in Fig. 4a.

The reference spectra provided by the ORTEC GLP HPGe detector (Fig. 4b) does not show

Transition	Transition energy for various elements (keV)			
	μC	μO	μAl	μAr
K_α	75.258*	133.535	346.828	644.004
K_β	89.212*	158.422*	412.877†	770.6†
K_γ	94.095*	167.125*	435.981†	815.0†
L_α			65.756*	126.237
L_β			88.771*	170.420*
L_γ			99.423*	190.870*

Table 1: Atomic transitions detected for different muonic atoms. The asterisk * indicates transitions that were not well resolved because of detectors limited energy resolution, while dagger † refers to those lines with small statistic or low efficiency of the detectors.

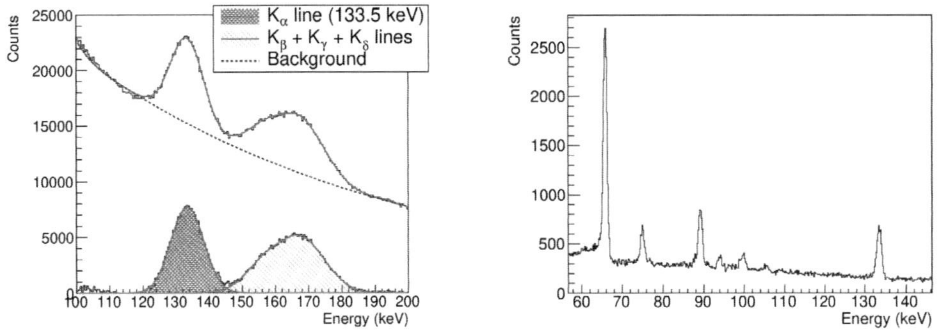

(a) LaBr$_3$ counters spectrum

(b) HPGe spectrum

Figure 4: X-ray spectrum recorded using **(a)** the four smaller LaBr$_3$(Ce) counters and **(b)** the HPGe detector.

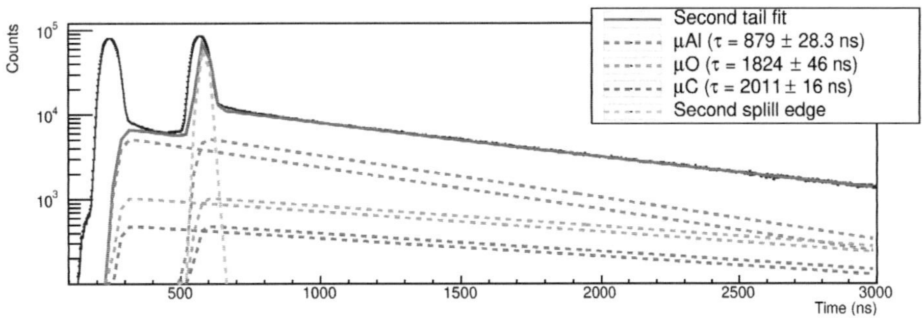

Figure 5: Time distribution of the events recorded with the H_2–CO_2 target using the four smaller LaBr$_3$ detectors.

characteristic X-rays due to unwanted contaminations, confirming the purity of the gas mixture under study.

As mentioned in the previous section, also the time distribution of the events needs to be reconstructed. From Fig. 5 it can be inferred that the time spectrum can be divided in two main contributions: the "prompt" peak, roughly in time with the muon spills, and the "tails" of the delayed events. The main part of the characteristic X-rays is concentrated inside the prompt peak, while the tails are populated by the products of muon decay (low energy electrons and bremsstrahlung photons); thus, while characteristic X-rays indicate the composition of the target, the tails of the distribution carry information about the muonic atoms lifetimes.

	This work (ns)	Suzuki et al. (ns)
μC	2011 ± 16	2026 ± 1.5
μp	2141 ± 98	2194.53 ± 0.11
μAl	879 ± 28	864 ± 2
μO	1824 ± 46	1795 ± 2
μAr	564 ± 14	537 ± 32

Table 2: Lifetimes of various muonic atoms as measured by the FAMU experiment and their values reported in [16] (and references therein).

Starting from the simplest target, the graphite block, the tails of the time spectrum have been fitted with gaussians convolved with exponentially decaying functions. The fit was performed considering only the far delayed part of the time spectrum ($t > 2500\,\text{ns}$) and the falling edge of the second muon beam spill, because in gaseous targets the time spectrum might present distortions in the time region near the muon spill caused by the muon transfer process.

The lifetimes obtained for the various muonic atoms are shown in Table 2 together with the values reported in literature [16]; the fit for the H_2–CO_2 and H_2–Ar targets was performed fixing the lifetimes of μp, μAl and μC using the values obtained from the graphite and the pure H_2 targets.

From these two gas mixtures, one can notice that the μp component of the time spectrum is strongly suppressed; this happens because all the μp atoms transfer their muons to O, C and Ar before they have time do decay.

4. Conclusions

The experimental method adopted by the FAMU experiment for the measurement of the proton Zemach radius requires a detection system suited for time resolved X-ray spectroscopy with high resolution both on energy and time. Scintillating detectors based on $LaBr_3$(Ce) crystals with their output recorded by a 500 MHz digitizer for an off-line pulse analysis were tested at the RIKEN-RAL muon facility and met the requirements both on energy and time resolution, correctly reconstructing the relevant characteristic X-rays and the lifetimes of the muonic atoms.

The good performance achieved with this detection techniques make $LaBr_3$(Ce) crystals optimal candidates for further studies on the muon transfer process and for the final experiment.

References

[1] Mohr P J, Newell D B and Taylor B N 2015 Codata recommended values of the fundamental physical constants: 2014 URL `http://dx.doi.org/10.5281/zenodo.22826`
[2] Pohl R *et al.* 2010 *Nature* **466**
[3] Antognini A *et al.* 2013 *Science* **339**
[4] Pohl R, Gilman R, Miller G A and Pachucki K 2013 *Ann. Rev. Nucl. Part. S.* **63**
[5] Zemach A C 1956 *Phys. Rev.* **104**
[6] Bakalov D, Milotti E, Rizzo C, Vacchi A and Zavattini E 1993 *Phys. Lett. A* **172**
[7] Bakalov D, Adamczak A, Stoychev L and Vacchi A 2012 *Nucl. Instr. Meth. Phys. Res. B* **281**
[8] Bakalov D, Adamczak A, Bakalova K, Polacco E and Rizzo C 2001 *Hyperfine Interact.* **136-137**
[9] Bakalov D, Adamczak A, Stoilov M and Vacchi A 2015 *Phys. Lett. A* **379**
[10] Mulhauser F and Schneuwly H 1993 *Hyperfine Interact.* **82**
[11] Werthmüller A, Adamczak A, Jacot-Guillarmod R, Mulhauser F, Schaller L, Schellenberg L, Schneuwly H, Thalmann Y and Tresch S 1998 *Hyperfine Interact.* **116**
[12] Jacot-Guillarmod R, Mulhauser F, Piller C, Schaller L A, Schellenberg L, Schneuwly H, Thalmann Y A, Tresch S, Werthmüller A and Adamczak A 1997 *Phys. Rev. A* **55**
[13] Jacot-Guillarmod R, Mulhauser F, Piller C and Schneuwly H 1990 *Phys. Rev. Lett.* **65**
[14] Matsuzaki T, Ishida K, Nagamine K, Watanabe I, Eaton G and Williams W 2001 *Nucl. Inst. Meth. Phys. Res. A* **465**
[15] Guffanti D 2015 *The FAMU experiment: measurement of muonic atoms spectra* Master's thesis University of Insubria
[16] Suzuki T, Measday D F and Roalsvig J P 1987 *Phys. Rev. C* **35**

Simulations of the Long Baseline Neutrino Experiment for the Sieroszowice Underground Laboratory (SUNLAB)

Małgorzata Harańczyk[1]

[1]Institute of Nuclear Physics Polish Academy of Sciences
Radzikowskiego 152, 31-342 Kraków, Poland

E-mail: malgorzata.haranczyk@ifj.edu.pl

Abstract. The Sieroszowice Underground Laboratory in Poland, SUNLAB, had been studied in the years 2008-2011 within the framework of the FP7 LAGUNA design study as an option for the realization of a next generation large volume neutrino observatory in Europe. However, in order to fully understand its physics capabilities, the feasibility studies of the SUNLAB laboratory have continued after 2011, including sensitivity calculations focused on the delta CP measurement for a large LArTPC detector at a distance of 950 km from CERN in a long baseline neutrino experiment. For this purpose the neutrino beam based on the SPS proton accelerator at CERN was simulated and the LAr data used to simulate the detector response.

1. Introduction

Neutrino studies have entered a very exciting era. The discovery of neutrino mixing and oscillations over almost the two past decades has provided solid evidence for new physics beyond the Standard Model. Recently, θ_{13} has been determined to be moderately large ($\sin^2 2\theta_{13} = 0.084^{+/-} 0.005$ [1]). This discovery has opened up exciting prospects for current and future long baseline neutrino oscillation experiments towards the answers to the remaining fundamental questions, in particular the type of the neutrino mass hierarchy and the value of CP-violating phase. There is no doubt that a demonstration of CP violation in neutrino oscillations will give an important clue for models of leptonic CP violation and leptogenesis. The combined analysis from the currently running and foreseen for the near future experiments will very probably lead to the determination of the neutrino mass hierarchy. However, the measurement of CP phase is even more challenging and requires higher event statistics, i.e. high intensity neutrino and antineutrino beams and a large, sensitive detector. Several worldwide projects concerning long baseline neutrino experiments aiming at CP violation discovery have been recently discussed. The LBNF project located in the USA with the baseline of 1300 km distance, coupling the Fermilab - neutrino beam with the DUNE - far detector (Liquid Argon TPC) in the Homestake historical mine, is the most advanced one with granted financial support. Other projects, such as Hyper-Kamiokande with a 295 km baseline situated in Japan and the LAGUNA-LBNO project with the longest 2300km baseline from CERN to Pyhasalmi mine in Finland, are being or have been discussed.

The physics-oriented studies of the Sieroszowice Underground Laboratory (SUNLAB) as a long baseline neutrino experiment assume a 950 km long baseline and a neutrino beam from CERN -SPS to a far detector located in the Polkowice-Sieroszowice mine. The chosen technology for the far detector is the Liquid Argon Time Projection Chamber (LAr-TPC), known for its ability to provide excellent tracking and calorimetry performance.

2. Neutrino oscillations

Neutrino oscillation is a quantum mechanical phenomenon in which a neutrino changes its flavour as it travels. This phenomenon takes place if neutrinos have non-degenerate masses and if mixing occurs. For three neutrino flavours, ν_e, ν_μ, ν_τ, and three mass eigenstates, ν_1, ν_2, ν_3, the mixing matrix is defined by six independent parameters. There are two independent mass-squared differences among the three neutrino masses: $\Delta m_{21}^2 = m_2^2 - m_1^2$, responsible for solar neutrino oscillations, and $\Delta m_{31}^2 = m_3^2 - m_1^2$, responsible for atmospheric neutrino oscillations; three mixing angles: θ_{12} - solar mixing angle, θ_{23} - atmospheric mixing angle, θ_{13} - a mixing angle which connects the solar sector with the atmospheric one; and one Dirac-type CP phase δ_{CP}.

The neutrino oscillations are studied in two ways: in the appearance experiments, where the probability that a neutrino of flavour α from the source will arrive at the far detector as a neutrino of flavour β is determined (e.g. $P(\nu_\mu \rightarrow \nu_e)$), or in the disappearance experiments, where the probability of a neutrino surviving with the same flavour is measured (e.g. $P(\nu_\mu \rightarrow \nu_\mu)$). Depending on which oscillation parameter is to be investigated, the oscillation channel should be chosen so as to maximally emphasize the visible measurable effect.

The interesting contribution to the oscillation effect from CP violating part, proportional to $\sin(\delta_{CP})$ and $\sin(\theta_{13})$, can be probed in a neutrino oscillation experiment by measuring the appearance probability of an electron flavour. For the disappearance experiments ($\alpha=\beta$) the imaginary CP-violating term of the transition probability equation becomes zero.

During propagation through matter, the weak interaction couples the neutrinos to matter. The ordinary matter, here the Earth's mantle, consists of electrons, protons and neutrons but it does not contain any muons, tau-leptons or antiparticles, such as positrons. The neutrinos of all three flavours (ν_e, ν_μ and ν_τ) interact with the electrons, protons and neutrons of matter through flavour independent neutral current interaction mediated by Z bosons. Moreover, the electron neutrinos have an additional contribution because of their charged current interactions with the electrons which are mediated by the $W^\pm$ exchange. This matter contribution creates an additional difference between the neutrino and anti-neutrino oscillation probabilities, which is the larger the longer baseline is assumed. As a result, besides the CP asymmetry caused by the CP phase δ_{CP}, we have also a "fake" asymmetry induced by matter, which causes difficulties in extracting the information on δ_{CP}. However, this "fake" asymmetry is related to the yet unknown neutrino mass hierarchy.

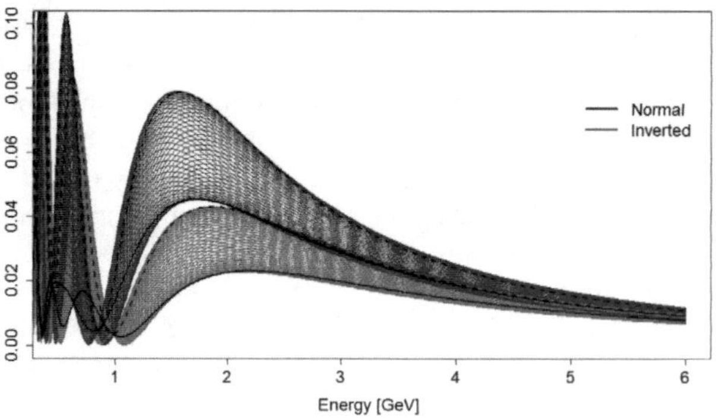

Figure 1 ν_e appearance probability as a function of neutrino energy for 950 km and δ_{CP} ($-\pi,\pi$); Normal mass hierarchy - blue, Inverted mass hierarchy - red.

Journal of Physics: Conference Series **689** (2016) 012019 doi:10.1088/1742-6596/689/1/012019

The sign of Δm_{31}^2 is unknown, hence two patterns of neutrino masses are possible: $m_3 > m_2 > m_1$, called the Normal Hierarchy, where Δm_{31}^2 is positive, and $m_2 > m_1 > m_3$, called the Inverted Hierarchy, where Δm_{31}^2 is negative.
The global fit [2] of all the available neutrino oscillation data in three-flavour analysis gives as the best fit for mass differences: $\Delta m_{21}^2 = 7.5 \times 10^{-5}$ eV^2 and $|\Delta m_{31}^2| = 2.4 \times 10^{-3}$ eV^2 with the relative 1σ precision of 2.4% and 2.8%, respectively; for mixing angles: $\sin^2\theta_{12} = 0.3$ (1σ precision is 4%), $\sin^2\theta_{13}=0.023$ (1σ precision is 10%), $\sin^2\theta_{23}= 0.41$(1σ precision is 12%); CP phase δ_{CP} is unknown. These values were used in the neutrino oscillation simulations presented below. Figure 1 shows the calculated probability of the electron neutrino appearance after travelling a distance of 950 km through Earth for two mass hierarchies: Normal and Inverted, for δ_{CP} values from the range $(-\pi,\pi)$ - each thin line represents different δ_{CP} value. For a better illustration, the maximal effect from CP phase $\delta_{CP} = \pi/2$ (black solid line) and $\delta_{CP} = -\pi/2$ (black dashed line) is depicted.

3. The SUNLAB laboratory

The idea of SUNLAB was developed in 2007, when seven sites in Europe were considered as candidate hosts of the next generation large volume neutrino observatory within the LAGUNA FP7 design study. The location in the Polkowice-Sieroszowice mine in Poland was discussed among them and the study on excavation of large detector caverns was performed [3]. Independently, a project of a small low background underground laboratory was considered and included in the Polish Roadmap of Research Infrastructures in 2015. The SUNLAB small laboratory will be placed 950 meters below the surface in a salt-rock characterized by an extremely low level of natural radioactivity. The results of the measurements of natural radioactivity in the Sieroszowice salt-rock strata can be found in [4], while more recent measurements performed using the Germanium gamma spectrometers in the underground salt caverns are given in [5].

The Polkowice–Sieroszowice mine belongs to the KGHM Polska Miedź S.A. holding of copper mines (with additions of other metals and salt layers as well) and is located in Kaźmierzów, Polkowice district, Lower Silesia province in south-western Poland. Geologically, it is located in the so called Zechstein formation, a unit of sedimentary rock layers.

4. Neutrino beam simulation

A conventional neutrino beam is produced from a decay of secondary mesons (mainly pions and kaons), which are created by high energy primary protons hitting a thick target. The direction of the neutrino beam is defined by the decay tunnel towards the far detector site. The magnetic horns mounted just behind the target make it possible to choose positively (or negatively for an antineutrino beam) charged mesons of a desired energy and direct them into the decay tunnel. In our simulations, using the Geant4 package [6,7], the SPS at CERN is assumed as the source of 400 GeV protons. It is assumed to deliver 1.5×10^{20} protons on target per year. The graphite target is 1 m long and two magnetic horns are placed in the so called NuMI design (horn and reflector). This kind of a setup can be operated in two modes by switching the direction of the circulating current inside the horns: positive horn focusing (PHF) mode for neutrino beam and negative horn focusing (NHF) mode for antineutrino beam. By changing the size and position of the elements of the beamline setup, one can optimise the expected neutrino flux. In our case, the first oscillation maximum for the 950 km baseline is expected at neutrino energies around 1.95 GeV (Figure 1) and the simulated setup has been optimised accordingly. An example of the optimisation process is shown in Figure 2 for the expected neutrino and antineutrino flux for three target positions. By moving the graphite target inside the horn (downstream - green line) the mesons of lower energy are chosen, and by placing the target more upstream (blue line) we get a wider spectrum and with higher energy neutrinos. In our studies, the most optimal position for the target was found just at the entrance of the horn, in the so called zero position (yellow line).

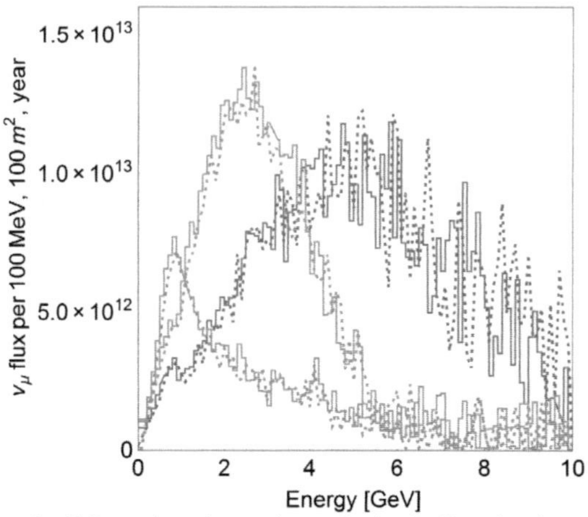

Figure 2 The muon neutrino (solid) and antineutrino (dashed) flux in the case of three different target vs horn positions

The conventional muon neutrino beam from mesons decays contains contaminations by other neutrino flavours. Pions and kaons decay into muons and muon neutrinos, while the muon itself produces a muon antineutrino, electron neutrino and positron while decaying. Also kaons may decay into an electron, electron neutrino and pion. These additional neutrino flavours are the main source of background for the experiments studying the electron-neutrino appearance in the muon neutrino beam. By carefully choosing the length of the decay tunnel one can minimise the number of muon decay processes. In our simulation the length of the decay tunnel is 100 m. The final simulated neutrino flux with all neutrino flavours for both PHF and NHF modes is presented in Figure 3.

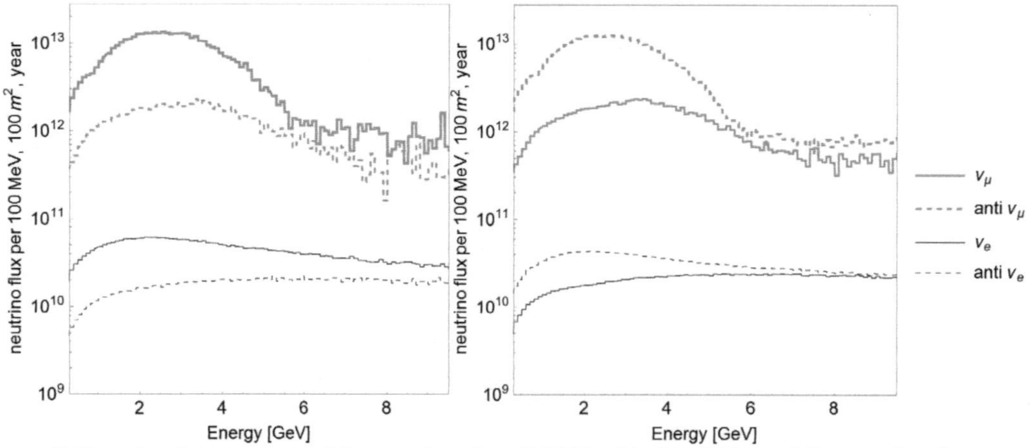

Figure 3 Neutrino beam composition produced at CERN with the SPS used for accelerating protons. The neutrino flux at a distance of 100 km per year. The PHF (left) and NHF (right) modes.

5. Far detector parameterisation

In this study the Liquid Argon Time Projection Chamber is assumed as a far detector located in SUNLAB. LAr TPC can provide excellent imaging with a good energy resolution for large active volumes. This allows us to include in the analysis the events with complicated topology. This technology was successfully tested during the operation of the ICARUS-T600 experiment in Gran Sasso [8] and currently is under vivid development. For this simulation two detector masses are assumed, the first phase detector - 20 kton of liquid Argon, and the large final phase detector - 100 kton.

The reconstructed neutrino energy threshold is assumed to be 300 MeV. The detection efficiency for charged current electron (ν_eCC) and muon (ν_μCC) events is 100%. The energy resolution of the detector is expected to be good and here it is assumed to be 15% of $\sqrt{E/GeV}$ for ν_eCC events and 20% of $\sqrt{E/GeV}$ for ν_μCC. For the ν_e appearance signal the ν_eCC events in the far detector are searched for. The efficiency for the rejection of the background due to neutral current events is assumed to be 99.5%. An additional background source comes from muons wrongly identified as electrons - here less than 0.5%. The last assumed background comes from the intrinsic beam contamination of electron neutrinos - the most difficult to reduce, here assumed at the level of 20%.

6. Sensitivity of CP violation

The neutrino oscillations, expected event rates and sensitivity of the δ_{CP} measurement, assuming the LAr detector at SUNLAB, were calculated using the GLOBES software [9]. The Earth matter effects were included and the oscillation parameters were used as described in Section 2.

Figure 4 depicts the expected signal and background event spectra in the ν_e appearance channel as a function of the reconstructed neutrino energy, including the efficiency and the background rejection capabilities described in Section 5. A large influence of neutral current background events (green dashed line) in the low reconstructed energy region is visible. On the other hand, in the higher energy region the beam contamination by electron neutrinos (red dotted line) is dominant.

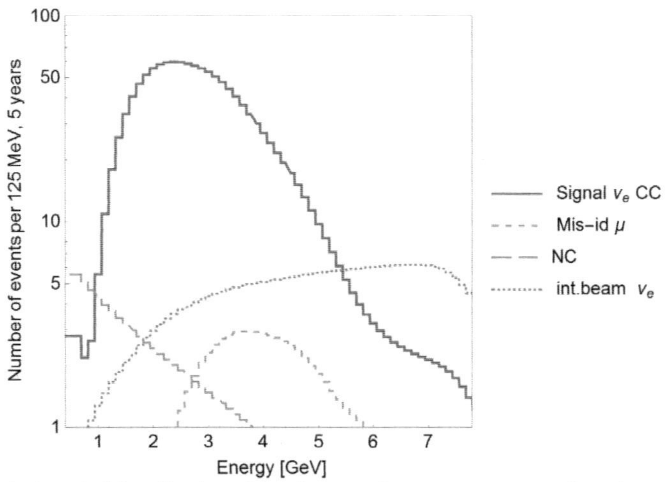

Figure 4 The expected signal (blue line) and background event rates as a function of the reconstructed neutrino energy for appearance channel. PHF mode for 5 years, 100 kton LAr TPC, Normal hierarchy, δ_{CP}=0 assumed.

The expected event rate calculation can be done for all possible δ_{CP} values and mass hierarchy scenarios. Based on this calculation, one can estimate the sensitivity of the CP violation discovery. The CP violation sensitivity is defined here as the ability of the experiment to distinguish between CP violation and CP conservation at an assumed confidence level, where the CP conservation refers to δ_{CP}=0 or π. The sensitivity to the CP violation (non vanishing δ_{CP}) is obtained by testing the hypothesis of the CP conservation evaluating the $\Delta\chi^2$ as a function of the value of δ_{CP}. The evaluated $\Delta\chi^2$ is plotted as a function of δ_{CP} test value assumed as an alternative hypothesis.

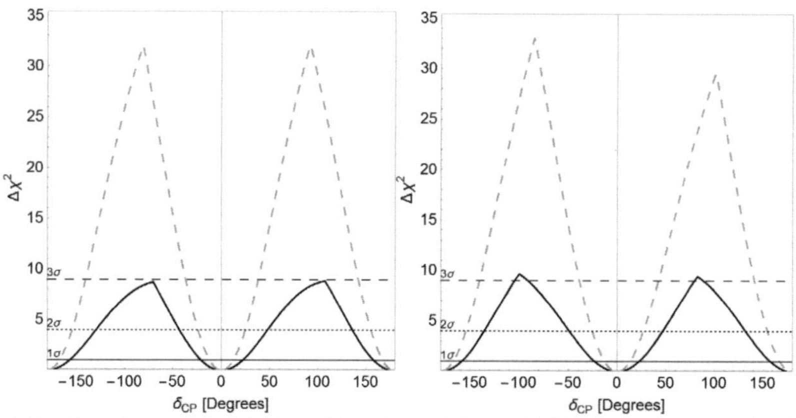

Figure 5 Sensitivity for the non-vanishing CP phase. Normal Hierarchy assumed - left, Inverted Hierarchy assumed - right.

Figure 5 shows the $\Delta\chi^2$ contours, which should be interpreted as the sensitivity to the CP violation. For this calculation the information from both neutrino and antineutrino modes were included and 5 years of data taking for each mode were assumed (10 years in total). The sensitivity was calculated separately for Normal and Inverted neutrino mass hierarchy. The black line shows the first phase scenario with the 20 kton LAr TPC detector, while the dashed orange line represents the final phase large detector of 100 kton. It is clearly visible that the SUNLAB based experiment will be most sensitive in the region of maximal CP violation (δ_{CP}=90 or -90 degrees) and in this case even the first phase detector can measure the effect on 3σ confidence level after 10 years of running. The smaller the δ_{CP} value, the more difficult the measurement, so a large 100 kton detector or a very long exposure is needed to measure the effect. These results are consistent with other neutrino long baseline projects.

7. Summary

Presented simulations show that the SUNLAB laboratory equipped with a large LAr TPC detector coupled with neutrino and antineutrino beams produced at CERN-SPS could measure the non-zero CP phase reasonably well.

The project of the small underground laboratory in the Polkowice - Sieroszowice mine will be unique in terms of an extremely low background from natural radioactivity.

Acknowledgments

We acknowledge the support from the Polish National Science Centre under the Preludium grant DEC-2011/03/N/ST2/01971.

References

[1] An F P, et al., *Phys. Rev. Lett.* **108**, 171803
[2] Gonzalez-Garcia M, Maltoni M, Salvado J and Schwetz T 2012 Global fit to three neutrino mixing: critical look at present precision *J. High Energy Phys.* JHEP12(2012)123
[3] Zalewska A, et al. 2010 *Acta Phys. Pol. B* **41**, 1803
[4] Kisiel J, et al. 2010 *Acta Phys. Pol. B* **41**, 1813
[5] Polaczek-Grelik K, et al. 2015 *J. Radional. Nucl Ch.* doi:10.1007/s10967-015-4567-6
[6] Longhin A 2010 *AIP Conf. Proc.* **1222** 339-343
[7] Agostinelli S, et al. 2003 *Nucl. Instrum. Meth. A* **506** 250-303
[8] Rubbia C, et al., 2011 *J. Instrum.* **6** P07011
[9] Huber P, Lindner M, Winter W 2005 *Comput. Phys. Commun.* **167** 195

The LArIAT Experiment at Fermilab

Irene Nutini[1,2,3], on behalf of the LArIAT Collaboration
[1]Università degli Studi di Firenze - Piazza di San Marco, 4, 50121 Firenze, Italy
[2]Fermi National Accelerator Laboratory, U.S. Department of Energy - Neutrino Division
[3]GSSI - Gran Sasso Science Institute, Viale F. Crispi, 7, 67100 L'Aquila, Italy

E-mail: irene.nutini@gssi.infn.it

Abstract. The LArIAT experiment at Fermilab is part of the International Neutrino program recently approved in the US. LArIAT aims to measure the main features of charged particles interactions in argon in the energy range (0.2 - 2.0 GeV) corresponding to the energy spectrum of the same particles when produced in a neutrino-argon interaction (neutrino energies of few GeV) typical of the short- and long-baseline neutrino beams of the Neutrino Program.
Data collected from the 1st Run are being analyzed for both Physics studies and a technical characterization of the scintillation light collection system. Two analysis topics are reported: the method developed for charged pion cross section measurement, based on the specific features of the LArTPC, and the development and test of the LArIAT custom-designed cold front-end electronics for SiPM devices to collect LAr scintillation light.

1. Introduction

The LArIAT experiment stands for Liquid Argon (Time Projection Chamber) In A Testbeam [1, 2].

The main purpose of the experiment is to perform a precise calibration of the Liquid Argon TPC detector. The LArIAT experiment is focused on the study of all the charged particles that could emerge from (ν,Ar) interaction, for neutrino energies E_ν of few GeV, typical of the SBN and LBN programs. Therefore the detector is placed on a dedicated, low momentum TestBeam line at Fermilab (FTBF) which provides a controlled environment to produce and select particles of several species, momentum and charge (mainly pions $\pi^{\pm}$, protons p, muons $\mu^{\pm}$, electrons $e^{\pm}$, kaons $K^{\pm}$...). The main goals are the characterization of the signal from different particles in the LArTPC and the optimization of the offline algorithms for Particle IDentification and calorimetric reconstruction for liquid argon detector technology, both TPC wires signals and LAr scintillation light collection.

The LArIAT detector combines the "traditional" Liquid Argon TPC (LArTPC) design for the ionization track imaging [3] with a new scintillation light collection system (LCS) providing enhanced light collection efficiency and uniformity [4]. The LArIAT TPC and the vacuum-insulated cryostat in which it is hosted are pre-existing components from the ArgoNeuT experiment [5]. The LArTPC consists of a rectangular box structure in which anode wires are tensioned at one side while at the other side we find the cathode plane. It has an active volume of 47 cm (width) x 40 cm (height) x 90 cm (length, along beam axis), corresponding to 170 litres of LAr (@ 90 K). There are three wire planes to see the drifting electrons inside the TPC. Induction and collection planes wires are read with dedicated cold electronics. Pulse signals are processed

Journal of Physics: Conference Series **689** (2016) 012020 doi:10.1088/1742-6596/689/1/012020

and analyzed through LArSoft, a dedicated framework for liquid argon TPC tracks reconstruction. The 3-D reconstruction of the event (tracks and energy deposit) in the TPC is made by matching the hits/clusters on the two wire planes having the same drift time in the two views. The LArIAT light collection system consists of an array of two high quantum efficiency cryogenic photomultiplier tubes (PMT) and three Silicon PhotoMultiplier Detectors (SiPM). They are deployed in liquid argon and mounted behind the wire planes of the TPC. Since the light collection is performed with these optical devices Very-UltraViolet LAr scintillation photons are wavelength-shifted into optical (blue) ones by covering the boundaries of the TPC volume with TPB coated reflective foils.
A schematic representation of the LArIAT assembled system is shown in Fig. 1. In Fig. 2 the LArIAT cryostat placed along the beamline is reported.

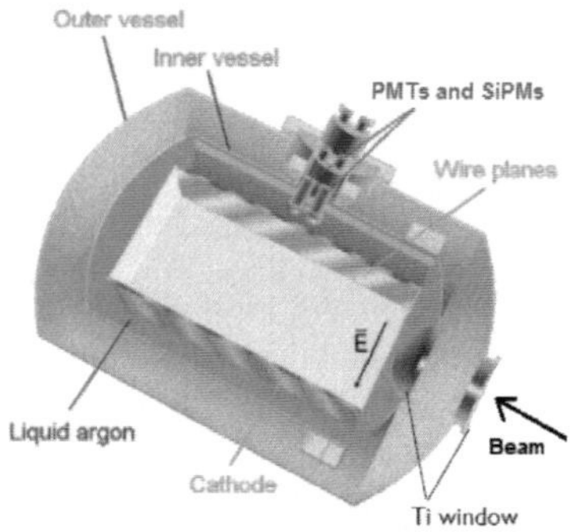

Figure 1. Schematic representation of the LArIAT detector [6].

Figure 2. LArIAT cryostat hosting the TPC placed in the beamline at FTBF - MCenter [7].

The 1st Runs of the LArIAT experiment lasted from April, 30th to July, 7th 2015. The analysis of the collected data are on going both for Physics studies (pion interaction total and exclusive cross sections, kaon interactions and decays, muon/pion separation) and for a technical characterization of the LCS - the new scintillation light collection system implemented in LArIAT. Some preliminary results have been shown in the poster presented at YRM2015 and are here briefly discussed.

2. LArIAT goal: Charged pion on Ar total cross section measurement

Charged pions are produced in a noticeable number in neutrino interactions with nuclei, like Ar, for neutrino energies of few GeV. They have very large interaction cross sections with nuclei, especially nearby the Δ resonance energy region. This is the reason why pions have been chosen as one of the main components of the charged particle TestBeam for the LArIAT experiment.

One of the goals of the LArIAT experiment is the experimental measurement of charged pion cross section on Ar in the (0.2 - 2.0) GeV energy range. Actually there are no experimental measurements for pions on Ar in this energy range. Current MonteCarlo (MC) simulation codes, like Geant4, use interaction models for Ar based on extrapolations from data with lighter or heavier elements. The goal of a dedicated pion run with LArIAT is to develop pion identification algorithms based on their interaction modes in argon and to exploit direct and precise measurement of the pion-nucleus cross section. This in turn will reduce the uncertainty on the hadron interaction models adopted in MC simulations for Ar target when dealing with neutrino interaction products (neutrino energy reconstruction and event topology recognition in future experiments with LAr detectors).

Journal of Physics: Conference Series **689** (2016) 012020 doi:10.1088/1742-6596/689/1/012020

Let us consider the pion-nucleus total cross section σ_{tot} for a pion traveling through a target. In general the target is not a single particle, but a slab of material containing many diffusion centers. If we assume the target centers uniformly distributed and the target thin enough not to have one center sitting in front of another one, we are in the so called *thin target* approximation. To first order, we will consider the pion in the nucleus interacting with individual "free" nucleons either p or n. To perform an experimental measurement of the σ_{tot} dependence on energy, we aim to scan different pion initial kinetic energies (N_{inc}, number of pions shot onto the target) and to count how many pions have experienced an interaction (N_{int}) and how many of them have crossed the thin target without interacting (N_{surv}). From the rate of interactions at each energy we can calculate the total cross section σ_{tot} from:

$$N_{int}(x, E) = (1 - N_{inc}e^{-\sigma_{tot}(E)Nx}) \tag{1}$$

where x is the target thickness (in cm), along the incident pion direction, N is the scattering center density in the target $N = \frac{\rho N_A}{A}$ (in cm^{-3}) and and σ_{tot} is the total cross section per nucleon (in cm^2).

When the "thin target" approach is not applicable, i.e. in the case of thick target experiments, to get a precise evaluation of the pion cross section dependence on energy, it is necessary to know the energy lost by the pion while crossing the target before the interaction, i.e the energy of the primary particle at the interaction point. For this reason we have to take into account the primary particle energy deposition by collisions with atoms of the medium before the interaction. When the target is thin, it is reasonable to consider only the pion-nucleus hadronic interactions, while for thick target experiments we should also consider the possibility of pion decay (in flight and at rest) in the target volume and the capture at rest on target nuclei for negative pions π^- at the end of range (near stopping).

The LArIAT TPC active volume is equivalent to a "thick target" for charged pion interactions. To estimate the charged pion cross section on Ar from the data acquired in LArIAT, we developed a new offline analysis method, the "Sliced TPC" approach, that makes use of the fine granularity and the high spatial resolution of the detector to "slice" the thick TPC target volume into many thin target LAr slabs [7]. The two wire planes of the TPC are composed by 240 wires oriented at +/- 60° at 4 mm distance. Each wire thus collects signals ($\propto$ released energy) in a 60° inclined 4 mm thin slab of Liquid Ar. It is rather "natural" to subdivide the TPC volume in 240 slices of $\Delta x = 4\ \mathrm{mm}/\sin(60°) \approx 5$ mm thickness along the x axis, i.e. direction of the incident particle (pions). Each slice n can be now considered as a "thin target" and we can apply the cross section calculation from Eq. 1 iteratively, evaluating the actual kinetic energy of the particle at that slab E_n^{kin}:

$$E_n^{kin} = E_{inc}^{kin} - \Sigma_{i=1}^{n}(\frac{dE}{dx})_i \Delta x \tag{2}$$

The results here reported come from MC simulation studies of charged pion interactions in a liquid argon target. I did a comparison at MC simulation level between ($\pi^{\pm}$, Ar) cross section results in a thin target and in a thick target, the last treated with the method previously described.

A prediction of the total hadronic interaction ($\pi^{\pm}$,Ar) cross section dependence on the kinetic energy of the pion has been produced from "thin target" simulations with two different MC generators (Geant4 [version 10.1] using FTFP_BERT physics list - Bertini Cascade model [8, 9] and Genie [version 2.8.2] using Intranuke hA model). It appears that the two MC generators pretty well reproduce the cross section at high energy for several targets, while they give slightly different predictions for the total cross section in the Δ resonance energy region. We think that the simulation models for elastic interactions need to be better tuned. The "thick target" simulation was performed with Geant4 "at particle level", i.e. no other detector features were taken into account and simulated, except for the slices geometry in the LAr TPC volume. At each step/slice the MC returns the kinetic energy of the pion entering the slice and the information whether an interaction occurred in the slice or not, the same information it is possible to extract from the energy deposition along the track for real events. Using Eq. 1 for a target thickness x = 1 cm (two slices), the pion cross section $\sigma_{tot}(E)$ is finally obtained for several energies.

The comparison between cross section results in thin and thick target with Geant4 has validated the "Sliced TPC" method for a high accuracy and statistically precise measurement of the ($\pi^{\pm}$, Ar) cross section with the LArTPC from the LArIAT experiment. In Fig. 3 the total cross section MC results for π^- are shown. The ongoing analysis on LArIAT collected data deals with the reconstruction and selection of charged pion events, the analysis of the calorimetric information along the track in the TPC, the hadronic interactions tagging and vertex finding and the evaluation of the total cross section applying the "Sliced TPC" method. The systematic errors are going to be estimated by varying the track reconstruction and pion tagging parameters both at the MC level and on the data.

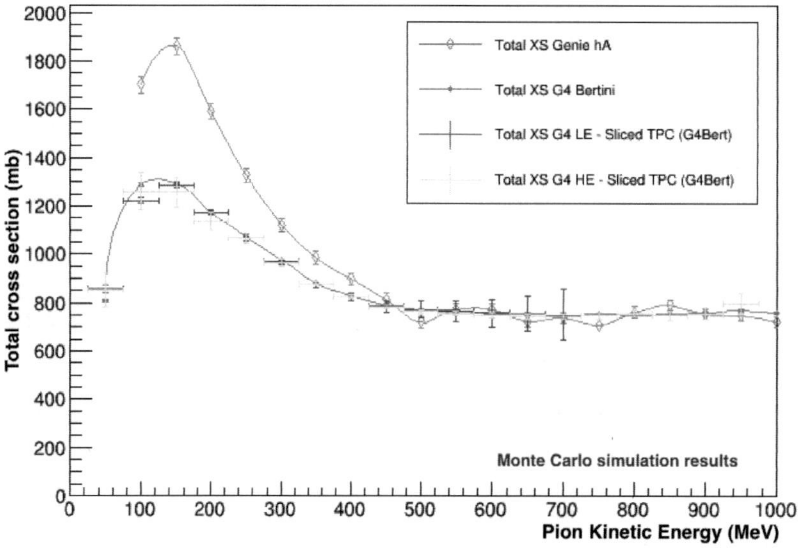

Figure 3. π^- on Ar total cross section dependence on the kinetic energy from MC simulations: comparison between Geant4 and Genie predictions for a LAr thin target and extrapolated cross section with the "Sliced TPC" approach with a Geant4 Bertini based simulation of the LArTPC thick target [7].

3. LArIAT goal: LAr scintillaton light collection with SiPMs with cold front-end electronics

One more technical goal for the LArIAT experiment is the development, test and characterization of dedicated devices for LAr scintillation light collection. In addition to the collection of the ionization charge from the TPC, the detection of the scintillation light produced in LAr is an actual point of interest, both for its trigger function for the TPC events collection and for the possibility of using also this information to improve the energy resolution of this detector technology. Actually two different light collection systems have been implemented in the LArIAT cryostat: two PMTs and three SiPM readout boards. The development and test of LAr light collection optical devices, especially dedicated cold front-end electronics for SiPM devices, for LArIAT will be here briefly reported.

The common approach for SiPM front-end electronics consists in reading the SiPM signals directly (without a preamplification stage) or with a preamplification stage outside the detector (or outside the cryostat, for cryogenic detectors), using warm electronics [10]. We decided to follow a new different way, reading these signals with front-end electronics (amplification-integration step) directly deployed in LAr working at cryogenic temperatures. Since there is a great interest and effort in SiPM commercial production and development, we decided to use and test devices from different producers in LArIAT: two Hamamatsu SiPM MPPC Arrays S11828-3344M and one single channel SiPM SensL MicroFB 60035 [11]. We designed, produced and tested (both at room temperature, 300 K, and in LN_2, 77 K) three different preamplification boards with dedicated cold electronics for the three SiPMs. Then we mounted the boards on the cryostat optical flange and deployed them in LAr behind the TPC wire planes, as shown in Fig.4.

Figure 4. PMTs and SiPMs boards mounted on the inner flange of the LArIAT cryostat [7].

A brief comment about the preliminary characterization of the SensL SiPM (0.36 cm^2 active surface, coupled with OPA656 cold amplifier [12]) response to LAr scintillation light produced by cosmic rays in the TPC volume is here reported.

The SiPM preamplified signal is higher in amplitude and faster than the direct signal and reproduces multiple and single photoelectron pulses. The average waveform for signals collected with similar trigger conditions (e.g. cosmic trigger) has a shape comparable with the LAr scintillation time distribution and its area corresponds to the average number of photoelectrons collected by the SiPM. Studying the single photoelectrons pulses due to LAr late light in the signal tails we obtained the SER (Single Electron Response) plot from which we had an estimate of the Gain of the readout setup in LAr for a fixed bias voltage. A preliminary estimate of the SiPM Light Yield (LY) in the LArTPC has then been evaluated: $LY_{SensL} = 0.12$ pe/MeV. This value is consistent with the predictions from the comparison with the PMT LY and with the light collection simulations.

A more complete characterization of all the three SiPMs response in LAr and the comparison with the PMT results is planned for these next months. Moreover the ongoing development for light collection for LArIAT 2nd Run is related to the possibility of making SiPM matrices that could cover the same area of a PMT photocathode. We designed and produced four cold-readout boards for SensL SiPM arrays. Tests and characterization of the response of these boards in cryogenics are ongoing.

4. Conclusions

The LArIAT collaboration effort is focused on the data analysis from the 1st Run, both for Physics studies (e.g charged pion total cross section on Ar, ...) and for technical improvements (SiPM cold readout boards tests and characterization, ...). Results are expected to improve particle ID and event reconstruction in liquid argon while also contributing to useful physics that will aid analyses in other LArTPC experiments, as MicroBooNE, SBND, DUNE...

The 2nd Run will begin in February 2016 after some minor changes (cryogenic technical improvements, anode wires electronics modifications, installation of new larger SiPMs mounted to custom readout boards...).

References

[1] Cavanna F, Kordosky M, Raaf J, Rebel B and LArIAT collaboration, 2014, LArIAT: Liquid Argon In A Testbeam, arXiv:1406.5560 [physics.ins-det]
[2] LArIAT/T-1034: Liquid Argon TPC In A Test beam, TSW - Technical Scope of Work for the 2013 FERMILAB Test Beam Facility Program, 2013
[3] Rubbia C, 1977, The Liquid-Argon Time Projection Chamber: a new concept for neutrino detectors, *CERN EP Internal Report* pp 77-8
[4] Szelc A M, 2013, The LArIAT light readout system, *JINST* **8** C09011.
[5] Anderson C et al, 2012, The ArgoNeuT detector in the NuMI low-energy beam line at Fermilab, *JINST* **7** P10019
[6] Rebel B and LArIAT collaboration, LArIAT presentation at FNAL Director's R&D Review, 2014, Internal documents
[7] Nutini I, 2015, Study of charged particles interaction processes on Ar in the 0.2 - 2.0 GeV energy range through combined information from ionization free charge and scintillation light., FERMILAB-MASTERS-2015-03
[8] Agostinelli S (GEANT4 Collaboration), 2003, Geant4 - a simulation toolkit, *Nucl. Instrum. Meth.* A **506** pp 250-303
[9] Heikkinen A, Stepanov N, Wellisch J P, 2003, Bertini intra-nuclear cascade implementation in Geant4, *CHEP 2003*, pp 24-28
[10] Collazuol G, 2012, The SiPM Physics and Technology- a Review, LAL Orsay
[11] SensL Single Channel SiPM MicroFB 60035, datasheet
[12] OPA656, Wideband, Unity-Gain Stable, FET-Input Operational Amplifier, Texas Instruments, datasheet

Measurement of cosmogenic radioactive products in xenon and copper

Francesco Piastra[1]

[1]Department of Physics, University of Zürich, Winterthurerstrasse 190, CH-8057, Zürich, Switzerland

E-mail: fpiastra@physik.uzh.ch

Abstract. Rare events searches, such as direct dark matter detection or neutrinoless double beta decay ($0\nu\beta\beta$) observation, using liquid xenon as target and detection medium require ultra-low background to fully exploit the physics potential. Cosmogenic activation of the detector components, and even more importantly, of the xenon itself might have undesired impact on the background and the final sensitivity of the experiment. Since no measurement of cosmogenic activation of xenon was present in literature so far, we performed such a measurement exposing of a natural xenon sample to the cosmic radiation at the Jungfraujoch research station at an altitude of 3470 m above sea level for 245 days. This study was complemented with a ultra pure copper sample that was activated together with the xenon. We directly observed, with gamma-ray spectrometry, the production of ^{7}Be, ^{101}Rh, ^{125}Sb, ^{126}I and ^{127}Xe in xenon, out of which only ^{125}Sb could potentially lead to a background relevant for multi-ton scale direct dark matter search. The production rates for five out of eight radioactive isotopes in copper are in good agreement with the only dedicated measurement present in literature. The production rates measured for both samples were compared with the predictions obtained with commonly used software packages. The latter showed a systematic under-estimation, especially for xenon.

1. Introduction

Particle detection based on liquid Xe (LXe) was widely developed in the last decade mainly in view of direct dark matter searches [1–5].
The expected spectrum of nuclear recoils from the elastic scattering of a weakly interacting massive particle (WIMP) off a nucleus has a quasi-exponential shape [6, 7]. Hence, the low energy threshold of such an experiment can heavily affect the detection sensitivity, especially for low mass WIMPs ($\lesssim$ GeV/c^2). Among the particle detection media, LXe features one of the highest scintillation and ionization yield [8,9] making this material one of the best candidates for dark matter searches.

Background from radioactive isotopes (radiogenic) and from cosmic rays (cosmogenic) may severely limit the detection sensitivity of any experiment searching for rare events. The former are present as residuals in every detector's materials or dissolved in the LXe itself (intrinsic sources), while the latter can be suppressed only operating the detector in deep underground laboratories, where the nuclear component of the cosmic rays (CR) is absent and the muonic component is suppressed by several orders of magnitude. The relatively high atomic number (Z=54) and density ($\rho \simeq 2.9\,\mathrm{g/cm^3}$) of LXe provide a good self-shielding power to γ and β rays able to efficiently suppress the radiogenic background in the most internal regions of the detec-

Table 1. Potential double β emitters of natural xenon. The half-lives of the double-beta decay isotopes ^{124}Xe, ^{134}Xe and ^{136}Xe are from the NuDat 2.6 database [16], the range for the double electron capture isotope ^{126}Xe is a theoretical prediction from [17]. The natural isotopic abundances are from [18].

Xe mass number	124	126	134	136
Abundance [%]	0.09(1)	0.09(1)	10.44(10)	8.87(16)
$T_{1/2}$ [yr]	$>1.6\times10^{14}$	$[5\text{-}12]\times10^{25}$	$>5.8\times10^{22}$	2.165×10^{21}

tor's sensitive volume. However this is not anymore true for intrinsic backgrounds such as ^{85}Kr and ^{222}Rn, that must be reduced to few ppt and few μBq, respectively. Moreover natural Xe has only few very long-lived isotopes that decay with double beta mechanism, listed in Table 1. Out of the potential double beta emitters, listed in Table 1, only the ^{136}Xe was observed so far and is currently used as source isotope for $0\nu\beta\beta$ searches [12–15].

The next generation ton and multi-ton scale LXe based detectors, in construction or in design phase, aim to probe the WIMP-nucleon spin independent cross section to few $\times10^{-47}\,\mathrm{cm}^2$ [4] with $\sim 2\,\mathrm{yr}\times$ ton of exposure or even to few $\times10^{-48}\,\mathrm{cm}^2$ [10, 11] with $\sim 20\,\mathrm{yr}\times$ ton. For the next-to-next multi-ton scale detectors, already in advanced design phase [5], the goal is to be sensitive to the ultimate cross section of $\sim 2\times10^{-49}\,\mathrm{cm}^2$ (with exposures of $\sim 10^2\,\mathrm{yr}\times$ ton), where the coherent neutrino-nucleus scattering events, which produce the same kind of signals expected from WIMP scatters off nuclei, become the dominant irreducible background. In detectors with these mass scales the radiogenic γ and β backgrounds will be dominated by intrinsic sources as detailed in [5]. The Xe exposure to the CR, during the production, transportation and storage above ground might produce unstable long lived nuclei which could result in an additional intrinsic source of background. Hence, it is critical, in particular for the next generations dark matter search experiments, to be able to predict the Xe cosmogenic activation and understand the maximum tolerable exposure to the cosmic rays (CR).

Software packages, like ACTIVIA [19] and COSMO [20], are able to predict the activation of yield of various materials given their isotopic composition, the nuclear CR spectrum and the exposure time. Calculations of Xe cosmogenic activation were performed with both codes for the XENON100 experiment [1] and resulted in large tension with the measured background [21,22]. The main cause of the discrepancy was attributed to the lack of knowledge of the production cross sections of Xe, which are implemented in the codes using semi-empirical models developed by Silberberg and Tsao [23–25].

Here we report the first dedicated study of cosmogenic activation of Xe, recently published in [26], where a 2.04 kg research-grade xenon (impurities $<$ 10 ppm) with natural isotopic aboundances was exposed to cosmic rays for 245 day at an altitude of 3450 m at the "High Altitude Research Station Jungfraujoch" [27]. A 10.35 kg oxigen-free high conductivity (OFHC) copper sample was activated together with the xenon in order to provide a benchmark for our activation measurements and predictions. This sample came from the same high pure copper batch (impurities $<$ 100 ppm) used to build inner components of the XENON100 detector (sample 6 of Table 1 in [28]).

2. Measurement and analysis of the activated samples

Before the exposure the xenon was stored underground for more than 1.5 yr at the Laboratori Nazionali del Gran Sasso (LNGS, Italy), where we assume that no activation takes place. The xenon sample was contained in a 1 liter stainless steel bottle (m = 2.9 kg) at a pressure of $\sim$ 100 bar ($\overline{T} \sim 20\,^{\circ}$C). In order to avoid the measurement results being dominated by the activation of the stainless steel bottle we used two identical bottles, one always kept underground and used for the γ measurements and the other used to expose the xenon to the CR. The gas transfer between the bottles was performed in the underground laboratory before and after the activation.

The nuclear activation processes in atmosphere are dominated by the high-energy nucleons, while the atmospheric muons play only a sub-dominant role because of their lower flux [29] and their smaller cross section [30, 31]. In order to compare our measurements with the predictions from the software packages, the activation values at high-altitude must be converted to sea level production rates. For our calculations we used the CR spectrum taken from [32, 33] (sea level values) scaled for the atmospheric depth according to [34]. The atmospheric depth at the activation site was determined by the use of the U.S. Standard Atmosphere 1976 model [35].

For both the samples, before and after the activation, γ spectra were acquired with the high purity germanium detector Gator [36], running at the LNGS underground laboratory. The background of this spectrometer is < 0.1 counts$\cdot$day^{-1}keV^{-1} over the entire energy region (from 50 to 2700 keV), and is sensitive to the contaminations of the most common γ emitters. The γ spectra of the xenon sample are shown in Figure 1 with the positions of the most relevant lines of the detected radioisotopes.

Before the activation the xenon sample was measured in the cavity for 26.5 days and the copper sample for 34.3 days. The pre-activation spectra were used to estimate the detection sensitivity to the γ-lines of the expected activation products as function of measuring time. The activation took place at 3450 m above the sea level, corresponding to an atmospheric depth of 728 g/cm^2, where the CR flux is 11.2 times higher with respect to the sea level. After the activation the samples were stored for 4 days at an altitude of 795 m at Lauterbrunnen, where the CR flux was 2.4 time lower than the previous location and the activation took place at a reduced rate. The considered cool-down time of the activated samples, from the time they were brought underground until the start of the measurement, was 2.5 days for the xenon and 14.8 for the copper. The γ-spectrum was acquired over 11.5 days for the xenon and 4.0 days for the copper.

The intrinsic activity from a radioisotope is inferred from the intensity of its most prominent γ-lines in the post-activation spectrum. In order to infer the activity of a radioisotope from several of its γ-lines at the same time the spectrum was analysed with a Bayesian method, described in [37–39]. The spectrum around each line is divided in three regions, a central signal region and two lateral control regions. The signal region is defined as the $\pm 3\sigma$ around the mean position of the peak, where σ is the energy-dependent resolution determined by proper calibrations of the spectrometer [36]. The background in the signal region is inferred interpolating the counting rates in the control regions, 3σ above and below the signal region. Hence the total counting rate expected in the signal region is given by:

$$\gamma_{\mathrm{S}} = m \cdot A \cdot (\varepsilon \cdot \mathrm{BR}) + w_{\mathrm{S}} \frac{\beta_{\mathrm{L}} + \beta_{\mathrm{R}}}{w_{\mathrm{L}} + w_{\mathrm{R}}}, \tag{1}$$

where m is the sample mass, A is the specific activity of the sample, $\beta_{\mathrm{L,R}}$ are the background rates in the left and right control regions, respectively, and $w_{\mathrm{S,L,R}}$ are the widths of the three regions. The product of branching ratio BR and detection efficiency ε of the γ-line is calculated with Monte Carlo simulations, based on a detailed implementation of the detector and sample geometry in GEANT4 [40]. The decays of the radio-isotopes are simulated using the G4RadioactiveDecay class, where branching ratios and directional correlations are taken into account.

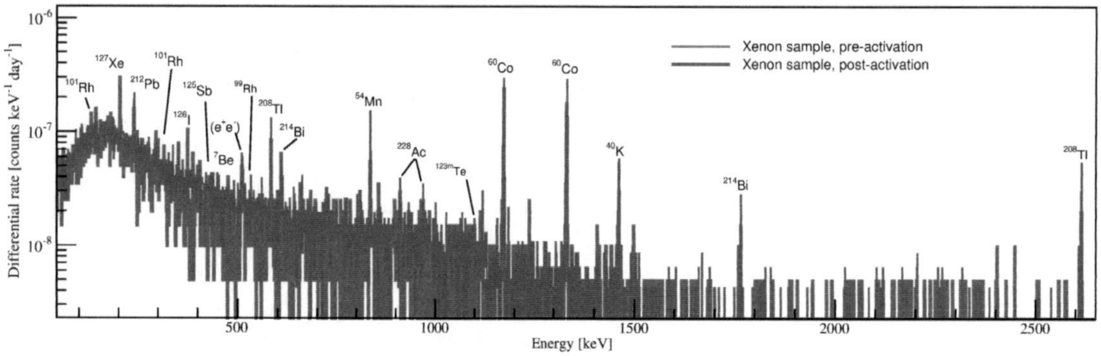

Figure 1. Pre- and post-activation spectra of the 2.04 kg xenon sample. One can identify the ^{127}Xe lines at 202.9 and 375.0 keV, and the ^{126}I line at 388.6 keV. Other prominent lines are from radioactive contaminations in the stainless steel bottle containing the xenon (primordial ^{238}U and ^{232}Th chains, ^{40}K, cosmogenic ^{54}Mn, ^{60}Co). Figure taken from [26].

Considering N γ-lines for the same isotope, the likelihood of the model is

$$\mathcal{L} = \prod_{k}^{N} f_P(C_{\mathrm{S}_k}|\gamma_{\mathrm{S}_k} t) \cdot f_P(C_{\mathrm{L}_k}|\beta_{\mathrm{L}_k} t) \cdot f_P(C_{\mathrm{R}_k}|\beta_{\mathrm{R}_k} t), \tag{2}$$

where the $C_{\mathrm{S}_k,\mathrm{L}_k,\mathrm{R}_k}$ are the counts in the three regions for each line k, and t is the measurement time. $f_P(C|\mu) = \mu^C/C!\ e^{-\mu}$ is the Poisson distribution.

The marginalized posterior probability density function (PDF) of the specific activity A is generated by Markov Chain Monte Carlo methods implemented in the Bayesian Analysis Toolkit (BAT) [41], where we use flat priors for the parameters γ_k and $\beta_{\mathrm{L}_k,\mathrm{R}_k}$. In Figure 2 some examples of marginalized posterior PDFs are shown for three different isotopes.

Since for some of the investigated isotopes the life time is comparable to the acquisition time, the specific activity A in Eq. 1 represents a mean specific activity over the measuring time. Hence the activity at the start of the acquisition, A_0 is:

$$A_0 = A \cdot \left(\frac{t}{\tau}\right)\left(1 - e^{-t/\tau}\right)^{-1}, \tag{3}$$

where τ is the radioisotope mean lifetime and t is the measurement time.
The specific activity of a sample of mass m after an activation time t is:

$$A(t) = \frac{N(t)}{m\ \tau} = P\left(1 - e^{-t/\tau}\right), \tag{4}$$

where P is the production rate which corresponds to the saturation activity $A(t \to \infty)$. Hence the production rate at the Jungfraujoch A_{J} is:

$$A_{\mathrm{J}} = A_0 e^{\left(\frac{t_{\mathrm{c}}+t_{\mathrm{t}}}{\tau}\right)}\left[\left(1 - e^{-t_{\mathrm{a}}/\tau}\right) + r\left(e^{t_{\mathrm{t}}/\tau} - 1\right)\right]^{-1}, \tag{5}$$

where t_{a} is the activation time at 3450 m, t_{t} is the activation time at 795 m, t_{c} is the cool-down time, and $r = P_{\mathrm{t}}/P_{\mathrm{J}}$ is the ratio of the production rates at Lauterbrunnen and Jungfraujoch, respectively.

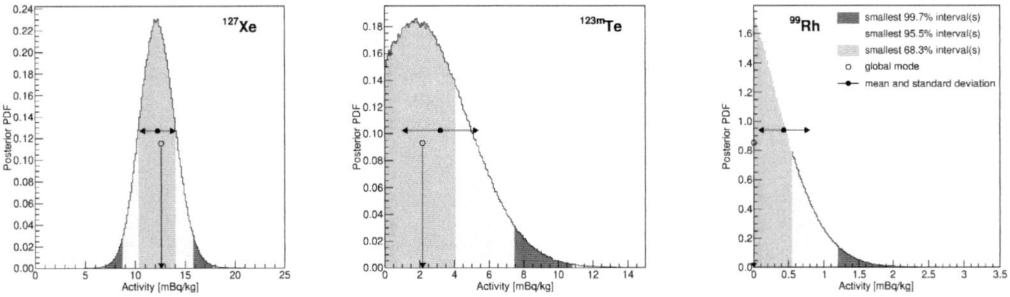

(a) (b) (c)

Figure 2. (a) Posterior PDF for the detected ^{127}Xe activity, clearly visible in Figure 1. The green region represents the shortest 68.3% C.I., which is taken as $\pm 1\,\sigma$ uncertainty. (b) For ^{123m}Te a signal is present, but an upper limit is reported since the intensity is too low to be determined. (c) An upper limit is reported for the activity of ^{99}Rh. The right edge of the yellow region represents the 95.5% upper limit (or the lower 95.5% C.I.). Figures extracted from [26].

3. Results and discussion

All the cosmogenic products of Cu that we identified are well known: they are spallation and fragmentation products of ^{63}Cu and ^{65}Cu, the only stable isotopes of copper. Most of these measurements (Table 2) are predicted within the 68 % credibility interval (CI) from the calculations performed with ACTIVIA and COSMO. The only exceptions are the measured activities of ^{57}Co and ^{58}Co that resulted $\sim$ 35 % and $\sim$ 20 % higher, respectively, with respect to the predictions. Furthermore five out of eight detected radioisotopes were found in agreement with the only other measurement available in literature so far, which was performed at LNGS [42] in 2009. In particular we observed production rates for ^{54}Mn, ^{57}Co and ^{60}Co lower by (2.5 ± 0.6), (1.7 ± 0.5) and (2.9 ± 0.8) times, respectively.

For the xenon sample we detected 5 cosmogenic products: ^{7}Be, ^{101}Rh, ^{125}Sb, ^{126}I and ^{127}Xe. The measured saturation activities are compared with the predictions with ACTIVIA and COSMO in Table 2. With the exception of ^{126}I, we observed a systematic underestimation of the production rates from the calculations performed with both packages. The observed production rate for ^{127}Xe was found in agreement within the 1σ CI with the measurement of the first 90 days of data reported by the LUX experiment ($1530 \pm 300\,\mu$Bq/kg) [43].

Among this detected radioactive products of xenon only the ^{101}Rh, ^{125}Sb have life times long enough to affect the sensitivity of an experiment foreseen to run for $\sim$ 5 years. We point out that both elements have a relatively high electronegativity and might be reduced by the xenon purification systems, based on hot zirconium getters, and commonly used in LXe detectors. However, this has not yet been proved experimentally.

The ^{101}Rh decays by electron capture to one of the excited states of the ^{101}Ru, which de-excite emitting only low energetic γ-rays in rapid sequence ($< 1\,$ns). This mechanism shifts, by pileup in LXe, the initial low energy recoils of the γ-rays to energies above the usual dark matter searching range.

On the other hand for the ^{125}Sb the situation is different as this isotope decays to ^{125}Te by β emission. In about 13.6 % of the decays the final state is the long-lived ($T_{1/2} = 57.4\,$d) metastable state at 145 keV, without any sbsequent γ-ray. Hence the low energy tail of these β decays would produce single-scatter electronic recoil background in the dark matter searching region, while all the other decay paths will not contribute to the background in the same energy region.

Table 2. Selected results, extracted from [26], for the cosmogenic activation of OFCH copper and natural xenon. The specific saturation activities at sea level are compared to our predictions based on Activia and Cosmo. The half-lifes refer to the numbers used in the software packages.

Specific saturation activity at sea level A_{sat} [μBq/kg]

Copper					Xenon				
Isotope	$T_{1/2}$ [days]	Measurement	Calculations Activia	Cosmo	Isotope	$T_{1/2}$ [days]	Measurement	Calculation Activia	Cosmo
^{46}Sc	83.79	27^{+11}_{-9}	36	17	^{7}Be	53.3	370^{+240}_{-230}	6.4	6.4
^{48}V	15.97	39^{+19}_{-15}	34	36	^{101}Rh	1205.3	1420^{+970}_{-850}	16.6	15.3
^{54}Mn	312.12	154^{+35}_{-34}	166	156	^{125}Sb	986.0	590^{+260}_{-230}	0.2	13.5
^{59}Fe	44.50	47^{+16}_{-14}	49	50	^{126}I	13.0	175^{+94}_{-87}	247	247
^{56}Co	77.24	108^{+14}_{-16}	101	81	^{127}Xe	36.4	1870^{+290}_{-270}	415	555
^{57}Co	271.74	519^{+100}_{-95}	376	350					
^{58}Co	70.86	798^{+62}_{-58}	656	632					
^{60}Co	1925.28	340^{+82}_{-68}	304	297					

4. Conclusions

We performed the first dedicated experimental study of the cosmogenic activation of xenon, employed as detection medium, and the second dedicated measurement of cosmogenic activation of OFHC copper, often used for the internal parts of low background experiments.

The agreement of the production rates measured for copper with the calculations performed with ACTIVIA and COSMO packages, supports the validity of the cosmic rays model we employed for the xenon study. However the very poor agreement of the calculations performed by the same packages for xenon suggests that experimental studies of the production cross sections are required in order to get reliable predictions for the cosmogenic activation of this material. Among the five cosmogenic products detected in xenon only ^{125}Sb might constitute a problem for the next generation dark matter detectors, while we did not observe any radionuclide that could be a problem for experiments looking for $0\nu\beta\beta$ decays of ^{136}Xe. However considering the electronic background published by the LUX experiment and using the activation times quoted for their xenon [43] the production rate we measured for ^{125}Sb seems to be too high. Thus we conclude that the activation rate is either close to the lower end of our quoted credibility interval, or ^{125}Sb is removed by the xenon purification system or by plating-out at the surfaces.

The only detected inert noble gas isotope, which may pose a problem for the next multi-ton scale detectors for WIMP searches, is the ^{127}Xe. However, given its relatively short half life ($T_{1/2} = 36.3\,\mathrm{d}$) the contamination from this isotope will be reduced already by an order of magnitude just after few months of storage underground.

As final remark we note that for the foreseen multi-ton scale LXe detectors, such as XENONnT [10], LZ [11] and DARWIN [5, 44], also the cosmogenic activation from the high energy muons fluxes in underground laboratories should be studied, as in this case the production in situ of short lived isotopes may contribute to the intrinsic background.

Acknowledgments

We gratefully acknowledge support from the staff and the International Foundation High Altitude Research Stations Jungfraujoch and Gornergrat (HFSJG) operating the Jungfraujoch laboratory [27], where the activation with cosmic rays was performed, the SNF grants 200020-149256 and 20AS21-136660, the ITN Invisibles (Marie Curie Actions, PITN-GA-2011-289442), Dr. D. Coderre, L. Bütikofer, and Dr. A.D. Ferella for help with transportation and handling the samples at LNGS, and Prof. Dr. C.J. Martoff for discussions on cosmogenic activation.

References

[1] Aprile E *et al.* (XENON Collaboration) 2012 *Astropart. Phys.* **35** 573
[2] Abe K *et al.* (XMASS Collaboration) 2013 *Nucl. Instr. Meth.* A **716** 78
[3] Akerib D S *et al.* (LUX Collaboration) 2013 *Nucl. Instr. Meth.* A **704** 111
[4] Aprile E *et al.* (XENON Collaboration) 2013, *Sources and Detection of Dark Matter and Dark Energy in the Universe. Proc. Symp. 10th UCLA DM2012, February 22-24, 2012* (Berlin: Springer) **148** 136 (*arXiv*:1206.6288)
[5] Schumann M *et al.* 2015 *JCAP* **10** 016 (*arXiv*:1506.08309)
[6] Jungman G *et al.* 1996 *Phys. Rept.* **267** 195
[7] Bertone G 2005 *Phys. Rept.* **405** 279
[8] Plante G *et al.* 2011 *Phys. Rev.* C **84** 045805
[9] Manalaysay A *et al.* 2010 *Rev. Sci. Instr.* **81** 073303
[10] Aprile E*et al.* (XENON Collaboration) 2014 *JINST* **9** P11006
[11] Malling D C*et al.* (LZ Collaboration) 2011 (*arXiv*:1110.0103)
[12] Auger M *et al.* (EXO Collaboration) 2012 *Phys. Rev. Lett.* **109** 032505
[13] Gando A *et al.* (KamLAND-Zen Collaboration) 2012 *Phys. Rev.* C **85** 045504
[14] Gando A *et al.* (KamLAND-Zen Collaboration) 2013 *Phys. Rev. Lett.* **110** 062502
[15] Martin Albo J *et al.* (NEXT Collaboration) 2013 *J. Phys.: Conf. Ser.* **460** 012010
[16] Nuclear Structure and Decay Database (NuDat 2.6), `http://www.nndc.bnl.gov/nudat2/`
[17] Shukla A *et al.* 2007 *J. Phys.* G*: Nucl. Phys.* **34** 549
[18] Rosman K J R and Taylor P D P (IUPAC) 1998 *Pure and Appl. Chem.* **70** 217
[19] Back J J and Ramachers Y A 2008 *Nucl. Instr. Meth.* A **586** 286
[20] Martoff C J and Lewin P D 1992 *Comput. Phys. Comm.* **72** 96
[21] Aprile E *et al.* (XENON Collaboration) 2011 *Phys. Rev.* D **83** 082001
[22] Kish A 2011 *PhD thesis* University of Zurich (`http://opac.nebis.ch/ediss/20121322.pdf`)
[23] Silberberg R and Tsao C H 1973 *Astrophys. J. Suppl.* 220 (I) **25** 315
[24] Silberberg R and Tsao C H 1973 *Astrophys. J. Suppl.* 220 (II) **25** 335
[25] Silberberg R *et al.* 1998 *Astrophys. J.* **501** 911
[26] Baudis L *et al.* 2015 *EPJ* C **75** 485
[27] `http://www.ifjungo.ch/jungfraujoch/`
[28] Aprile E *et al.* (XENON Collaboration) 2011 *Astropart. Phys.* **35** 43
[29] Beatty J J *et al.* 2004 *Phys. Rev.* D **70** 092005
[30] Heisinger B *et al.* 2002 Earth and Planetary Sci. Lett. **200** 345
[31] Hagner T *et al.* 2000 *Astropart. Phys.* **14** 33
[32] Ziegler J F 1996 *J. Res. Develop* **40** 19
[33] Stanev T 2010 *High Energy Cosmic Rays, 2nd*, Chichester UK, Praxis Publishing Ltd
[34] Beringer J *et al.* (Particle Data Group) 2012 *Phys. Rev.* D **86** 010001
[35] U.S. Standard Atmosphere 1976, `http://www.pdas.com/atmos.html`
[36] Baudis L *et al.* 2011 *JINST* **6** P08010
[37] Weise K and Matzke M 1989 *Nucl. Instr. Meth.* A **280** 103
[38] Weise K and Woger W 1992 *Meas. Sci. Techn.* **3** 1
[39] Lira I and Grientschnig D 2010 *Metrologia* **47** R1
[40] Agostinelli S *et al.* (GEANT4) 2003 *Nucl. Instr. Meth.* A **506** 250
[41] Caldwell A *et al.* 2009 *Comp. Phys. Comm.* **180** 2197
[42] Laubenstein M and Heusser G 2009 *Appl. Rad. Isot.* **67** 750
[43] Akerib D S *et al.* (LUX Collaboration) 2015 *Astropart. Phys.* **62** 33
[44] Baudis L *et al.* 2014 *JCAP* **01** 044

Journal of Physics: Conference Series **689** (2016) 012022 doi:10.1088/1742-6596/689/1/012022

The hybrid mesons quest: the MesonEx experiment at Jefferson Laboratory

Alessandro Rizzo[1] **for the CLAS Collaboration**

[1] The University of Rome "Tor Vergata" and INFN Roma2, Via della Ricerca Scientifica 1, 00133 Roma, Italy

E-mail: `alessandro.rizzo@roma2.infn.it`

Abstract. The meson spectroscopy plays nowadays a central role in the investigation of hadron structure thanks to the possible existence of exotic hybrid mesons, *quark-antiquark-gluon* bound states. Their explicit gluonic degrees of freedom which should clearly emerge from a Partial Wave Analysis (PWA) of the corresponding Dalitz plot of the exotic particle decay, may result in final J^{PC} configurations not allowed in the constituent quark model. Besides this clear signature, hybrid mesons are also expected to have a large particle multiplicity decays, requiring for their search an experimental apparatus with high performances in terms of rate capability, resolution and almost a full acceptance to apply PWA methods. New-generation experiments are planned at Thomas Jefferson National Laboratory (VA, USA) for which an unprecedented statistics of large multiplicity decay events with fully reconstructed kinematics will be available. In particular for the MesonEx (CLAS12) experiment in Hall B, a wide scientific program that will start in 2016 has been deployed to study the meson spectrum at energies up to 11 GeV. A key role in such program is played by the Forward Tagger apparatus of the experiment, which will allow to extend the study of meson electro-production to very low Q^2 values, in a quasi-real photo production kinematical region, where the production of hybrid mesons is expected to be favorite. Currently a new analysis framework for the search of the hybrid mesons is being set up by the HASPECT network, an international structure which gather people involved into theoretical and experimental hadronic physics all over the world. The goals of the network is to develop new analysis models and statistical techniques to unfold the signal and background distributions in high-statistics datasets. In this work are briefly presented the first preliminary results from the application of a statistical technique, namely the sPlot, to the data already acquired by the CLAS experiment for the decay $f_1(1285) \to (\eta)\pi^+\pi^-$.

1. The meson spectrum and the exotic hybrid configuration

The meson spectrum may carry fundamental information about the gluonic contribution in the hadron structure, thanks to the possible existence of states $q\bar{q}g$, exotic meson configurations not predicted by the *Gell-mann* model. These hybrid states are characterized by J^{PC} configurations not allowed for standard $q\bar{q}$ states[1], that may be experimentally investigated through Partial Wave Analysis (PWA) of the experimental data.
Several theoretical models include gluonic excitations contribution within different frameworks [1–3]: in this work the attention is focused on the Flux-tube model [4]. The gluonic flux tubes are

[1] J, P and C are the quantum numbers related to the total spin, the parity and the charge conjugation respectively, which univocally characterize the bound state.

a consequence of the non-abelianity of Quantum Chromo-Dynamics (QCD) theory, which, differently from Quantum Electro-Dynamics (QED), allows the gauge bosons (the gluons) to interact among each other, forming flux tubes which bind the interacting quarks. In this model, standard meson configurations correspond to a ground state flux tube which links the *quark-antiquark* pair, resulting in a J^{PC} configuration already predicted by the *Constituent Quark Model.* Considering the first excited states of the flux tube ($J^{PC} = 1^{+-}$ or $J^{PC} = 1^{-+}$) and arranging them with *pseudo-scalar* ($L = 0$, $s = 0$, $J = 0$) or *vector* ($L = 0$, $s = 1$, $J = 1$) mesons J^{PC} configurations, it is possible to form exotic hybrid configurations $J^{PC} = 0^{--}, 1^{-+}, 0^{+-}, 2^{+-}$, not predicted by the Gell-mann model and hybrid configurations $J^{PC} = 0^{++}, 1^{--}, 1^{++}, (0, 2)^{-+}, 1^{+-}$, already predicted by the Gell-mann model.
The flux-tube picture well explains also the hadronization process and the high particle multiplicity states in which a hybrid meson is supposed to decay in. A flux-tube binds two interacting quarks with a force of the form $\vec{F} = -k\vec{x}$, where k is a constant which is proportional to the strength of the interaction and $\vec{x}$ is the distance between the two quarks. The binding potential $V = -\frac{1}{2}kx^2$ grows rapidly with $|\vec{x}|$, till it reaches the energy of the mass of a quark pair. A this point the breaking of the flux-tube may occur with the formation of at least two new mesons in which the mother particle decays in.
In this picture the hybrid mesons are supposed to decay in high-particle multiplicity events because the flux-tube binding the quarks is in an excited state, a condition that enhances the energy available to form couples of new mesons.

2. Exotic hybrid mesons: hints and experimental evidences

Hints about the existence of such hybrid states come from Lattice QCD (LQCD) with unquenched calculations, that in these last years is approaching a pion mass considerably close to the physical one. The good predictive power of LQCD, due to the good agreement of lattice calculations and the experimental data for the meson spectrum in terms of number of states and mass hierarchy, makes the LQCD predictions about the existence of exotic hybrid mesons one of the pillars of the hybrid exotic mesons theory. In particular, according to these calculations, the lowest lying hybrid meson surprisingly presents the configuration $J^{PC} = 1^{-+}$ with a mass of about 1.6 GeV, followed by the configuration 0^{+-} with a mass of about 2 GeV: both states are lying in an energy range fully accessible to the CLAS12 experiment at Jefferson Laboratory.
The experimental history about evidences of such hybrid states starts in the 80's, when a joint CERN-IHEP experiment (GAMS group [5]) claimed for a possible existence of hybrid state $J^{PC} = 1^{-+}$: in the analysis of the channel $\pi^- p \to \pi^0 \eta n$, a p-wave contribution of the $\eta\pi$ bound state (namely $\pi_1(1400)$) was supposed to exist to better explain the behavior of the experimental data. To confirm that claim, the experiment E852 at the Brookhaven National Laboratory [6] started a series of systematic studies of the same channel and the crossed-charged one ($\pi^- p \to \pi^- \eta p$) at 18 GeV, observing the lower lying hybrid state. Also the experiments Crystal Barrel [7] and VES [8] confirmed such observations, by studying the channels $n\bar{p} \to \pi^- \pi^0 \eta$, $p\bar{p} \to \pi^0 \pi^0 \eta$ and $\pi^- N \to \pi^- \eta N$ respectively.
Another possible candidate for the lowest lying hybrid state $J^{PC} = 1^{-+}$ is the $\pi_1(1600)$, whose experimental evidence has been more controversial. The E851 experiment at Brookhaven National Laboratory claimed for the existence of such excited state [9] analyzing a dataset of 250k $\pi^- p \to \pi^- \pi^- \pi^+ p$ events at 100 GeV. In a following re-analysis of a higher statistical sample the previous claim was not confirmed by the same collaboration [10], but recently the hybrid $\pi_1(1600)$ has been observed with high statistics by the COMPASS experiment in the $\pi Pb \to \pi\pi\pi Pb$ channel [11].

3. The CLAS12 apparatus at the CEBAF accelerator: the meson spectroscopy program and the experimental investigation of hybrid mesons

The experimental investigation of hybrid mesons will be one of the main tasks of the spectroscopy program deployed for the MesonEx (or CLAS12) experiment, located at the experimental Hall B of the *Thomas Jefferson National Laboratory* (Virginia, USA). The recent upgrade of the Continuos Electron Beam Accelerator Facility (CEBAF) to 12 GeV [12] will allow MesonEx to investigate the mass range where the exotic states are supposed to lie. The accelerator is based on the Superconductive Radio-Frequency technology with two linacs, made of 25 cryo-modules each, placed between two sets of ricirculating arcs: such configuration enhances the energy of the accelerated electron beam by 1.2 GeV at each round-pass. When the beam reaches an energy of 11 GeV (5th pass) it is extracted to the Hall B, toward CLAS12 apparatus, with a maximum delivered current of 200 μA.
MesonEx (CLAS12) is the new experimental apparatus, a spectrometer, specifically designed for the upgraded CEBAF electron beam [13], that is replacing the former CLAS experiment, which took data till 2012 at energies up to 6 GeV. The geometry of the new apparatus ensures an acceptance which is close to 4π and a good hermeticity: it is divided into 6 sectors corresponding to 6 different azimuthal regions, and it covers both the regions around the target (Central Detector region) and the forward direction region (Forward Detector region). The magnetic field, which is used to measure the particle momenta, is generated by a superconductive solenoid magnet (5 T) in the Central Detector and by a superconductive torus magnet (3.6 T) in the Forward Detector. The different detectors of the experimental apparatus are described in detail in [13] and schematically in [14].
The CLAS12 experiment will look for exotic-hybrid mesons produced in quasi-real photoproduction reactions induced by the electron-beam from CEBAF impinging on nuclear target (LH_2) placed in the Central Detector region. The production of exotic hybrid mesons is expected indeed to be favored using a photon beam rather than a meson beam (K or π), because of the different final J^{PC} configurations accessible by a $J = 1$ probe (the photon) interacting with a nucleon. Recalling the considerations exposed in Section 1 we may see that arranging the first excited states of the flux tube ($J^{PC} = 1^{+-}$ or $J^{PC} = 1^{-+}$) with the J^{PC} configurations of vector mesons gives more exotic hybrid configurations ($J^{PC} = 1^{-+}, 0^{-+}, 2^{-+}$) than the one obtained by taking the pseudo-scalar mesons configurations ($J^{PC} = 0^{--}$). Moreover it is possible to obtain the lowest lying exotic hybrid states ($J^{PC} = 1^{-+}, 0^{-+}$) only considering the vector mesons quantum numbers: in the case of photoproduction the quantum number of the vector mesons ($J = 1$) is brought by the photon, so we expect such channel to be more favorably interested by the exotic-hybrid mesons production than the production via K or π beams ($J = 0$).

4. MesonEx and photoproduction reactions in Hall B

The Hall B at Jefferson Laboratory, where the CLAS12 (and previously CLAS) is located, has a longstanding experience about photoproduction reactions and experimental techniques used to obtain a photon beam from an electron one. During the 6 GeV era, for the CLAS experiment, the bremsstrahlung emission was used to produce a tagged real γ-ray beam [16]. In this technique the electron beam from CEBAF impinges on a thin diamond radiator, producing linear polarized photons via coherent bremsstrahlung emission: the electron and the produced photon beams are then separated by passing through a dipole which bends the electrons towards a scintillating *hodoscope*. Thanks to the energy conservation relation it is possible to associate the correct energy to the interacting photon in the nuclear target (E_γ) according to the formula $E_\gamma = E_e - E_{e'}$, where $E_{e'}$ is the energy of the deflected electron measured by the hodoscope and E_e is the electron beam energy. Nowadays this technique is used to produce the photon beam for the GlueX experiment in Hall D at Jefferson Laboratory, which, as the CLAS12 experiment, has an important meson spectroscopy program.

Journal of Physics: Conference Series **689** (2016) 012022 doi:10.1088/1742-6596/689/1/012022

In the 12 GeV era a different technique will be used for the CLAS12 experiment to study photo-induced reactions: the quasi-real photoproduction. Such technique is quite well established in higher energy physics experiments like HERMES, BABAR [17] and COMPASS [11], and it will be used in Hall B to search for the hybrid meson states with a complementary approach to the GluEx experiment. The quasi-real photoproduction technique takes place in the electro-production regime: when the electron beam impinges on the nuclear target, the electrons are scattered by an angle θ. Events with a small θ are the ones where the exchanged photon between the probe and the nucleon of the target has a low Q^2 or, in other words, the γ is quasi- on mass shell. Such events are those of quasi-real photoproduction.
In CLAS12 experiments the electrons, scattered at small θ angles, are detected by the Forward Tagger (FT), an apparatus placed immediately after the nuclear target (between the Central and the Forward Detector region, see Section 3), covering the polar angle range from 2.5 to 4.4 degrees. The FT apparatus is composed by a tracker, based on Micromegas technology, to measure the angle θ and the polarization plane of electrons; a scintillation hodoscope, used as a veto for gammas, and an electromagnetic calorimeter made of $PbWO_4$ crystals, to measure the scattered electrons energy. In a photoproduction event, tagged by the Forward Tagger, the multi-particle hadronic decay is detected by the CLAS12 apparatus, which provides a high rate capability, almost a 4π acceptance and excellent Particle IDentification (PID). Quasi-real photons will have an energy spectrum between 6.5 and 10.5 GeV, a linear polarization degree that can be up to the 70% of the electron beam one and a high luminosity, of the order of 10^{35} $cm^{-2}s^1$ (considering 5 cm long LH_2 target).

5. Establishing a new analysis framework to search for exotic hybrid mesons: the case of $f_1(1285)$ meson decay

The exotic hybrid mesons are expected to have a clear signature in a Partial Wave Analysis (PWA) of the experimental data, thanks to their *unconventional* J^{PC} configuration. The PWA is a powerful tool to search for this new physics but it requires several features to obtain reliable results, in particular high statistics, an almost full-acceptance of the experiment used to collect the data and a strong background rejection. In order to be ready for the first data-taking of the CLAS12 experiment in 2017, an international collaboration, namely HASPECT (HAdron SPEctroscopy CenTer), is currently working to set up a new analysis framework to search for new physics, in particular the lowest lying exotic hybrid mesons. During these years the collaboration studied if the CLAS12 experimental features will fulfill the requirements to successfully look for such states and, at the same time, developed new analysis tools. The excellent performances in terms of rate capability expected for the CLAS12 apparatus will fulfill the high-statistics requirements, while the apparatus acceptance and its influence on a possible PWA has been investigated with two complementary approaches. A first feasibility test has been done using simulated data for the $\gamma p \to \pi^+\pi^+\pi^- n$ channel where the sum of 8 isobar channels has been considered to describe resonances in S, P, D and F waves decaying into $\rho\pi$ or $f_2(1270)$. In this data an exotic contribution in the P wave has been injected to simulate the hybrid $\pi_1(1600)$ decaying in ρp, with a cross section of 200 nb. After folding the events with the MesonEx acceptance and processing them through the full reconstruction chain, the data have been fitted using a proper partial wave set in the framework of the isobar model, in selected bins of transferred momentum t. All known resonances, $a_2(1320)$ (D-wave), $a_1(1260)$ (D-wave), $\pi_2(1670)$ in the two decay channels ρp (P and F wave) and $f_2(1270\pi)$ (S and D wave), have been correctly reproduced at the right mass values, as well as the additional exotic signal. The second feasibility study is based on data already acquired by the CLAS experiment at 6 GeV. By studying the $\gamma p \to \pi^+\pi^-\pi^0 p$ benchmark reaction for the ω decay in three pions, the result has been projected on the CLAS12 acceptance and fitted. In this case it has been observed that the result is stable against the acceptance corrections, demonstrating that the PWA will

be feasible for the CLAS12 experiment.
The analysis of the decays of $f_1(1285)$ meson in the $(\eta)\pi^+\pi^-$ channel, concerns the development and the application of new analysis tools to reject the background. Using data already acquired by the CLAS experiment during the g11 run, in which a real photon beam was used to photo-produce the mesons, an analysis tools based on the sPlot method [19, 20] has been developed and tested on the Dalitz Plot of the decay (fig. 1). The sPlot [19] is a statistical method based

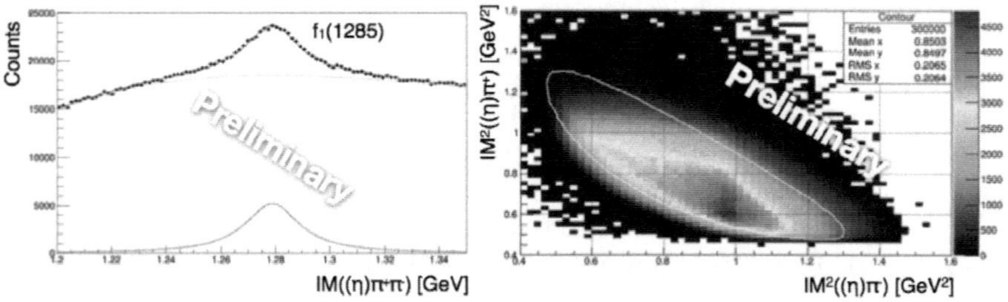

Figure 1: Left Panel: The invariant mass spectrum of the $(\eta)\pi^+\pi^-$ mesons and the PDFs used for the fit. In blue the Breit-Wigner function used to modeling the $f_1(1285)$ signal, in green the 4th grade Chebychev polynomial function used for the background and in red the sum of the two functions. Right Panel: Dalitz plot for the $f_1(1285)$ decay in terms of invariant mass square for the couples of mesons $(\eta)\pi^+$ versus $(\eta)\pi^-$. The yellow line is the theoretical Dalitz boundary.

on the maximum likelihood fit used to calculate a weight (namely the sWeight) for each decay event: the weight will have a different numerical value if the considered event is more likely background ($-1 < sWeight < 0$) or signal ($0 < sWeight < 1$).

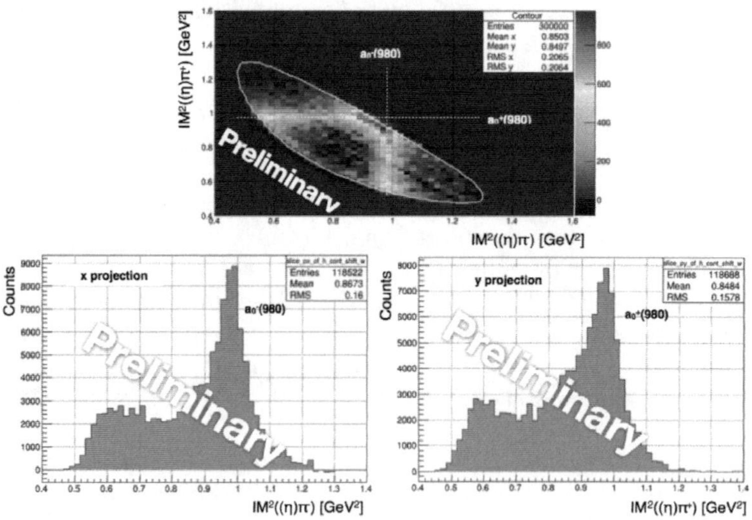

Figure 2: The sWeighted Dalitz plot of the $f_1(1285)$ meson decay in the $(\eta)\pi^+\pi^-$ channel. Even if the data are not yet acceptance corrected, it should be noted the presence of the "stripes" related to the $a^{\pm}(980)$ mesons, bound states of the $\eta\pi^+$ and $\eta\pi^-$ couple respectively

Each calculated sWeight is then associated to the corresponding event in the Dalitz, obtaining a Dalitz Plot with an enhanced signal over background ratio (see fig. 2). Without going too deep into technicalities of the method (see ref. [19]) we can review some basics of sPlot and its application to the $f_1(1285)$ decay Dalitz. The starting point is the spectrum of the invariant mass of $(\eta)\pi^+\pi^-$, (left panel), where both the non-resonant background contribution and the signal ($f_1(1285)$ mass) contribution are present. A Probability Density Function (PDF) is defined for the background shape F_b (*i.e.* a 4th Chebychev Polynomial) and for the signal F_s (*i.e.* a Breit-Wigner function) (see fig.1). A maximum likelihood fit is then performed on the $(\eta)\pi^+\pi^-$ invariant mass and the fitted functions F_s and F_b are used to evaluate the sWeight (W_s) according to the formula:

$$W_s = \frac{V_{ss}F_s + V_{sb}F_b}{N_sF_s + N_bF_b} \quad (1)$$

where the V_{ss} and V_{sb} are the covariance matrixes (defined in [20]), which are evaluated event by event, and N_s and N_b are the number of signal and background events respectively. By simply weighting by sWeight value each event in the Dalitz plot, univocally determined by the couple of values $(IM^2((\eta)\pi^+), IM^2((\eta)\pi^-))$, it is possible to obtain the background subtracted Dalitz plot as shown in fig. 2.

6. Conclusions

Considering the results reported in the previous section, it should be noted that even if the data are not yet acceptance-corrected, there is a sizable contribution of the two mesons $a_0^{\pm}(980)$ in the $f_1(1285)$ decay, whose contribution is generally quite elusive because of the quite huge background. In particular the projections of the sWeighted Dalitz shows a signal over background ratio which is of the order a factor five better and a statistics that is at least 70 times larger than the one available in literature [21–23]. The application of the sPlot method seems to be very effective and may be applied to data that will collected in the next years by the MesonEx experiment, allowing a PWA investigation to search for new physics. In particular the good performances in terms of rate capability, resolution and an almost full acceptance, makes MesonEx a privileged observation point to study the exotic hybrid mesons.

[1] Ketzer B 2012 arXiv 12085125 [hep-ex]
[2] Barnes T *et al.* 1995 arXiv 9501405 [hep-ph]
[3] Close F E and Page P R 1995 *Nucl. Phys. B* **443** pp 233-254
[4] Buisseret F *et al.* 2007 arXiv 0703020 [hep-ph]
[5] Alde D *et al.* 1988 *Phys. Lett. B* **205** 397
[6] Thompson D R *et al.* 1997 *Phys. Rev. Lett.* **79** 1630
[7] Abele A *et al.* 1998 *Phys. Lett. B* **423** 175
[8] Beladidze G M *et al.* 1993 *Phys. Lett. B* **313** 276
[9] Adams G S *et al.* 1998 *Phys. Rev. Lett.* **81** 5760
[10] Dzierba A *et al.* 2006 *Phys. Rev. D* **73** 072001
[11] Morréale A and the COMPASS collaboration 2011 *Journal of Physics Conference Series* **295** (1) 012110.
[12] Rode C H *et al. AIP Conf. Proc. 2010* **1218** 26
[13] Mecking B A *et al.* 2003 *Nucl. Instr. Meth. A* **503** 513
[14] Filippi A and the CLAS Collaboration 2015 *EPJ Web of Conferences* **96** 01013
[15] Szczepaniak A 1999 *Phys.Rev. D* **59** 034016.
[16] Sober D I *et al.* 2000 Nucl. *Instr. Meth. A 440* pp 263-284.
[17] Burkert V D 2006 Proceedings of XXII Symposium on Lepton and photon interaction at high energies, published by World Scientific, ISBN 981-256-662-7
[18] CLAS Collaboration 2012 CLAS12 Forward Tagger (FT) Technical Design Report, Technical Report available at `https://www.jlab.org/Hall-B/clas12_tdr.pdf`
[19] Pivk M and Le Diberder F R 2005 *Nucl. Inst. Meth. A* **555** pp 356-369
[20] Pivk M 2006 arXiv 0602023v1 [physics]
[21] Evangelista C *et al.* 1981 *Nuclear Physics B* **178** pp 197-241
[22] Teige S *et al.* 1998 *Phys. Rev. D* **59** 012001
[23] L3 Collaboration 2002 *Phys. Lett. B* **526** pp 287-294

Journal of Physics: Conference Series **689** (2016) 012023 doi:10.1088/1742-6596/689/1/012023

Probing ultrafast processes by fifth order Stimulated Raman Scattering

G Fumero[1], G Batignani[1,2], K E Dorfman[3], S Mukamel[3], T Scopigno[1,4]

[1] Dipartimento di Fisica, Università di Roma "La Sapienza", I-00185, Roma, Italy
[2] Dipartimento di Scienze Fisiche e Chimiche, Università degli Studi dell'Aquila, I-67100, L'Aquila, Italy
[3] Department of Chemistry, University of California, California 92697-2025, Irvine, USA
[4] Center for Life Nano Science @Sapienza, Istituto Italiano di Tecnologia, 295 Viale Regina Elena, I-00161, Roma, Italy

E-mail: giuseppe.fumero@gmail.com, tullio.scopigno@phys.uniroma1.it

Abstract. We present the full diagrammatic description of non-resonant impulsive femtosecond stimulated Raman spectroscopy in a multimode model system. In this technique the pump-probe scheme is exploited to study the vibrational structure of the sample via stimulated Raman scattering. We apply closed-time-path-loop diagrams to calculate the complete response of the system at the relevant perturbation order. We show that, in presence of low-frequency modes, coherences created by the impulsive pump modify the resulting Raman signal, which oscillates from gain to loss features, depending on the time delay between the pump and probe pulses. This leads to a redistribution of photons among the fields involved in the process and, consequently, the energy flows between fields and matter. Moreover, through this formalism, we address the case of extremely short delays in which the pump and probe fields overlap in time. We find that, even in absence of photo-induced dynamics due to absorption of the pump pulse, the overlap condition can generate time dependent features, arising from additional diagrams, which offer no contribution for well separated pulses.

1. Introduction

Ultrafast spectroscopy has recently emerged as a key tool in scientific inquiry, due to its capability to probe picosecond and sub-picosecond time scales at which atomic and molecular systems evolve. Revealing dynamics that last for extremely short amounts of time has a central role in the understanding of many fundamental and unknown phenomena that appear in numerous fields, from solid state physics to biochemistry [1–4].

The investigation of such dynamics, however, implies deciphering complex signals and requires a careful modeling to extract the information brought by the experimental measurements. The need for a powerful theoretical method to understand such signals has motivated the application of several perturbative techniques to spectroscopic problems and, in particular, to nonlinear spectroscopy [5]. Generally, nonlinear optical processes are described by the density matrix of the matter system interacting with light, while electromagnetic degrees of freedom are treated classically, through the Maxwell equations. The nonlinear signature of the process is contained in the induced polarization, which is calculated starting from the density matrix and acts as source in the Maxwell equations.

Specifically, a perturbative expansion in power of the fields leads the nth order polarization $P^{(n)}$, which describes $n+1$ light matter interactions, including the emission of the nonlinear signal. The full response is calculated summing all the possible orderings of the fields interactions and, consequently, all the possible pathways of the density matrix in its configuration space, called Liouville space. This is easily achieved through diagrammatic representations. Perturbation expansions based on the density matrix are represented by double-sided Feynman diagrams: the two sides of the diagram represent the ket and bra sides of the density matrix, while the arrows stand for excitations or de-excitations induced by the fields.
Another possibility consists in founding the perturbative expansion on the wave function of the complete system, coupling both the matter and the fields degrees of freedom quantum-mechanically. A different diagrammatic representation is then supplied by closed-time path-loop diagrams [6, 7], which describe the evolution of the wave function both forward and backwards in real time. This method allows for a dramatic reduction of the perturbative terms, since the number of loop diagrams scales linearly with n, while fully time ordered Feynman diagrams scale as n^2.
Here we apply loop diagrams to provide a full quantum description of non resonant Femtosecond Stimulated Raman Scattering (nr-FSRS) [8, 9]. This technique uses an ultrashort non resonant actinic pulse to create coherences of low-frequecy modes in the sample. The state of the system is then probed stimulating a Raman process through another mode of the system, using a narrowband Raman pulse and a broadband probe pulse. When the pump and probe pulses do not overlap in time, we find a modulation of the Raman signal, due to the actinic-induced low-frequecy coherences, in agreement with previous works [8, 10]. Additionally, our approach allows the microscopic tracking of the energy redistribution among fields and matter during the process. Moreover, we investigate the case of extremely short delay times between the pump and probe pulses. We demonstrate that in this instance additional terms have to be considered in the perturbative expansion.

2. The perturbative approach to non resonant FSRS

The nonlinear optical process underlying nr-FSRS is ruled by the Hamiltonian

$$H = H_0 + H', \tag{1}$$

where the unperturbed Hamiltonian H_0 is separable in a matter and an electromagnetic contribution, H_m and H_f respectively, and all its eigenfuctions are assumed to be known. The interaction term H' represents the effective radiation matter interaction. Assuming off-resonant excitations, the effective interaction Hamiltonian in the rotating wave approximation is given by

$$H'(t) = \alpha(t) \sum_{i,j} \mathcal{E}_j^\dagger(t)\mathcal{E}_i(t) + h.c. \tag{2}$$

where $\mathcal{E}_i$ and $\mathcal{E}_j^\dagger$ are the positive and negative frequency components for two different modes i and j of the field operator

$$\hat{E}(\mathbf{r}, t) = \sum_j \mathcal{E}_j(t)e^{i\mathbf{k_j}\cdot\mathbf{r}} + \sum_j \mathcal{E}_j^\dagger(t)e^{-i\mathbf{k_j}\cdot\mathbf{r}}. \tag{3}$$

In nr-FSRS experiments, the total field comes from the sum of the actinic pulse (E_a), Raman pump (E_p) and probe (E_s) fields. Since all the laser pulses are far from the resonance frequencies of the sample, excitations to virtual levels are ruled by the excited state polarizability operator

$$\alpha = \sum_{a \neq b} \alpha_{ab} \left|a\right\rangle \left\langle b\right|. \tag{4}$$

Although, as a matter of principle, the i and j run over all the possible modes of the total field operator $\hat{E}$, the energy and momentum conservation, as well as the experimental configuration of the pulses, limit this freedom and reduce the number of terms contributing to the signal in the summation. The stimulated signal is defined as the change in time of the mean number of photons $n_\omega(t)$ in the detected mode ω, belonging to the probe pulse E_s:

$$S = \int_{-\infty}^{+\infty} dt \frac{d}{dt} \langle n_\omega(t) \rangle = \int_{-\infty}^{+\infty} dt \frac{i}{\hbar} \Big\langle [H', n_\omega(t)] \Big\rangle = \frac{2}{\hbar} \int_{-\infty}^{\infty} dt \Im \Big\langle \alpha \mathcal{E}(t) \mathcal{E}^\dagger(t) \Big\rangle, \tag{5}$$

where $\langle n_\omega(t) \rangle = Tr[n_\omega(t)\rho(t)]$ and the time derivative has been computed through the Heisenberg equation of motion. Realistically, the signal is a convolution of the nonlinear response and the detector's response function. Thus, the integration in Eq. (5) is restricted by the acceptance width of the spectrometer. Taking advantage of the superoperators formalism in Liouville space [11, 12] and switching to the interaction picture, the stimulated optical signal obtained by a frequency gated measurement is given by [6]

$$S(\omega) = \frac{2}{\hbar} \Im \mathcal{E}_s^*(\omega) \int_{-\infty}^{+\infty} dt\, e^{i\omega t} \Big\langle \mathcal{T} P(t) e^{-\frac{i}{\hbar}\int_{-\infty}^{+\infty} dt'\, H'_-(t')} \Big\rangle, \tag{6}$$

where $\mathcal{T}$ is the time ordering superoperator, $P(t) = \alpha \sum_i \mathcal{E}_i^\dagger(t)$ is the induced polarization and H'_- is the Liouville superoperator in the interaction picture. We now introduce the perturbative step, expanding the exponential in Eq. (6) to the second order in α:

$$S(\omega) = (-1)^3 \Im \int_{-\infty}^{+\infty} e^{-i\omega t} \mathcal{E}_s^*(\omega) \mathcal{E}_{a/p}(t) \mathcal{T} \prod_{k=1}^{2} \int dt_k \sum_{i,j} \mathcal{E}_i(t_k) \mathcal{E}_j^*(t_k) \langle \alpha(t) \alpha(t_1) \alpha(t_2) \rangle . \tag{7}$$

This equation fully describes nr-FSRS and is a function of the fifth order nonlinear polarization $P^{(5)}$. The term $\mathcal{E}_{a/p}(t)$ represents alternatively a photon of the actinic or Raman pump pulse: indeed, a last excitation in the form $\alpha \mathcal{E}_s \mathcal{E}_s^\dagger$ leads to a negligible contribution, due to the spectral width of the femtosecond $\mathcal{E}_s$ pulse. Each term in the summation represents a different Liouville pathway and can be sketched by a loop diagram. The number of diagrams can be significantly reduced, considering energy conservation, phase matching and pulses configuration. Therefore, selecting the surviving diagrams simplifies the calculation of the nonlinear signal.

3. Loop diagrams selection

Loop diagrams represent the evolution of the wave function in the joint space of the matter and fields degrees of freedom. Field interactions are represented by pairs of arrows along the two branches of the loop. Interactions with the left or the right side of the loop account for a $\alpha \mathcal{E}_i \mathcal{E}_j^\dagger$ or a $\alpha^\dagger \mathcal{E}_i^\dagger \mathcal{E}_j$ term, respectively. Time flows from bottom to top, thus the two branches are not mutually time ordered, while interactions on the same branch are ordered.

We shall work in frequency domain, by Fourier transforming the field envelopes and defining retarded Green's function $G(\omega) = \frac{1}{\hbar} \sum_\alpha \frac{|\alpha\rangle\langle\alpha|}{\omega - \omega_\alpha + i\gamma}$ and the advanced Green's function $G^\dagger(\omega) = \frac{1}{\hbar} \sum_\alpha \frac{|\alpha\rangle\langle\alpha|}{\omega - \omega_\alpha - i\gamma}$ where $|\alpha\rangle$ is the matter eigenstate basis and γ is the vibrational dephasing rate. Thus, since the polarizability operator can be recast as $\alpha(\omega) = G(\omega)\alpha_{ab}$ where α_{ab} is the dipole constant associated to the transition $a \to b$, we can physically interpret free evolutions on the left and right sides of the diagrams as forwards and backwards in real time, respectively.

The nonlinear response is evaluated considering the possible bare diagrams and dressing them with the fields. Considering actinic excitations at frequencies lower than that of the Raman pulse and imposing energy conservation we select the relevant diagrams, represented in Fig. 1. Five diagrams contribute to the red side of the spectrum, with respect to the Raman pulse frequency, and five to the blue one. When the delay T between the actinic pulse $\mathcal{E}_a$ and the femtosecond probe pulse $\mathcal{E}_s$ is longer than the pulses duration, the first interaction is necessarily with the actinic field modes and the signal is represented by the four diagrams d_i-d_{iv}. When T is comparable to the femtosecond pulses duration, the nonlinear signal is also generated in the time region in which the three pulses overlap. In this case, the six additional terms d_1-d_6 arise from the summation in Eq. 1. By reading off the diagrams [13], we obtain:

$$S_n(\omega,T) = (-1)^3 P_g \Im \int_{-\infty}^{+\infty} \frac{d\Delta_p}{2\pi}\frac{d\omega'_p}{2\pi}\frac{d\Delta}{2\pi}\frac{d\Delta_a}{2\pi}\frac{d\omega'_a}{2\pi} 2\pi\,\delta(\Delta_p-\Delta_a-\Delta)e^{-i\Delta_a T}$$
$$\mathcal{E}_p(\omega'_p)\mathcal{E}^*_p(\omega'_p+\Delta_p)\mathcal{E}_s(\omega+\Delta)\mathcal{E}^*_s(\omega)\mathcal{E}_a(\omega'_a+\Delta_a)\mathcal{E}^*_a(\omega'_a)F_n(\omega,\Delta,\Delta_p,\omega'_p,\Delta_a,\omega'_a), \quad (8)$$

where the correlation functions F_n depend on the diagram as specified below. The thermal occupation of the state (i) is accounted for through a Boltzmann factor P_i, while the delta function provides energy conservation.

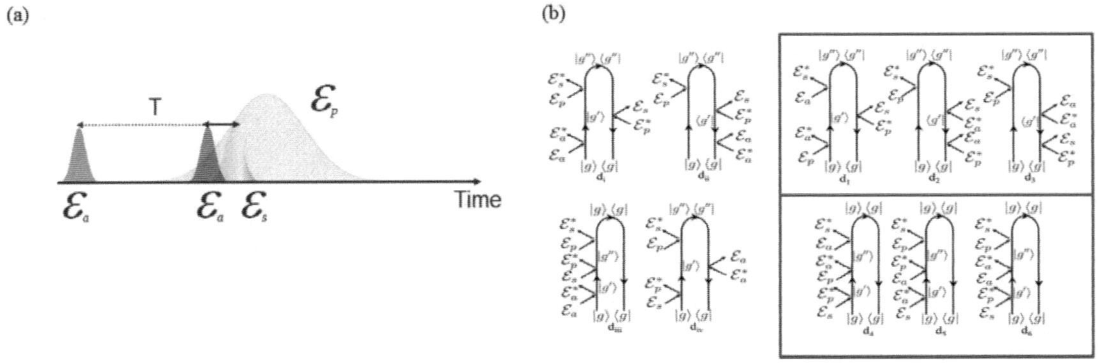

Figure 1: (a) Pulses configuration in nr-FSRS: the actinic pulse triggers a photoinduced process and precedes a couple of Raman pulse and Raman probe fields. When the delay T decreases, the two ultrashort pulses can eventually overlap in time. (b) The total fifth-order signal is represented by ten loop diagrams, five contributions (top panel) in the red side of the spectrum and five (bottom panel) on the blue side. Diagrams that contribute only within the overlap region are highlighted by a red box.

The correlation functions F_n contain the matter response; for diagrams d_i-d_6 we derive

$$F_{d_i} = \left\langle \alpha G^\dagger(\omega'_p-\omega+\Delta_a)\alpha G(\Delta_a)\alpha \right\rangle \qquad F_{d_{ii}} = \left\langle \alpha G^\dagger(\omega'_p-\omega)\alpha G^\dagger(\Delta-\Delta_p)\alpha \right\rangle$$
$$F_{d_{iii}} = \left\langle \alpha G(\Delta_a)\alpha G(\omega+\Delta-\omega'_p-\Delta_p+\Delta_a)\alpha \right\rangle \quad F_{d_{iv}} = \left\langle \alpha G^\dagger(\Delta-\Delta_p)\alpha G(\omega+\Delta-\omega'_p-\Delta_p)\alpha \right\rangle$$
$$F_{d_1} = \left\langle \alpha G^\dagger(\omega'_p-\omega'_a-\Delta_a)\alpha G(\omega'_p-\omega)\alpha \right\rangle \qquad F_{d_4} = \left\langle \alpha G(\omega-\omega'_a-\Delta_a)\alpha G(\omega-\omega'_p)\alpha \right\rangle$$
$$F_{d_2} = \left\langle \alpha G^\dagger(\omega'_p-\omega)\alpha G^\dagger(\omega'_p-\omega'_a-\Delta_a)\alpha \right\rangle \qquad F_{d_5} = \left\langle \alpha G(\omega-\omega'_p)\alpha G(\omega-\omega'_a-\Delta_a)\alpha \right\rangle$$
$$F_{d_3} = \left\langle \alpha G^\dagger(\omega'_p-\omega)\alpha G^\dagger(\omega'_p-\omega+\Delta_a)\alpha \right\rangle \qquad F_{d_6} = \left\langle \alpha G(\omega-\omega'_p)\alpha G(\omega-\omega'_p-\Delta_a)\alpha \right\rangle.$$

From the analysis of the correlation functions it emerges that diagrams d_i-d_{iv} and d_1-d_6 involve a different set of vibrational modes in order to generate non negligible signals. Remarkably, while

Journal of Physics: Conference Series **689** (2016) 012023 doi:10.1088/1742-6596/689/1/012023

diagrams d_i-d_{iv}, in addition to the probed mode, implicate a low-frequecy mode covered within the bandwidth of the actinic pulse, on the other hand diagrams d_1-d_6 require the presence of an extra level close to frequency difference between the Raman and actinic pulse.

4. Simulations for a model system

Simulations have been performed for a Gaussian actinic pulse centered at 650 nm and with a 50 fs FWHM and we assume a 490 nm monochromatic Raman pulse $\mathcal{E}_P(\omega) = 2\pi\, E_p\delta(\omega - \omega_P)$, which simplifies the frequency domain calculations.
Firstly, we consider a two level molecular system with a low-frequecy mode $\omega_L = 100\,\mathrm{cm}^{-1}$ and a mode $\omega_1 = 1400\,\mathrm{cm}^{-1}$. In this situation, only diagrams d_i-d_{iv}, contribute to the signal: low-frequecy coherences impulsively created by the actinic pump modify the resulting Raman signal, which oscillates from gain to loss features, depending on the time delay T. Depending on the specific pathway, this leads to the creation of main peaks at Raman frequencies as well as side peaks due to harmonic or anharmonic coupling between the modes involved in the process.

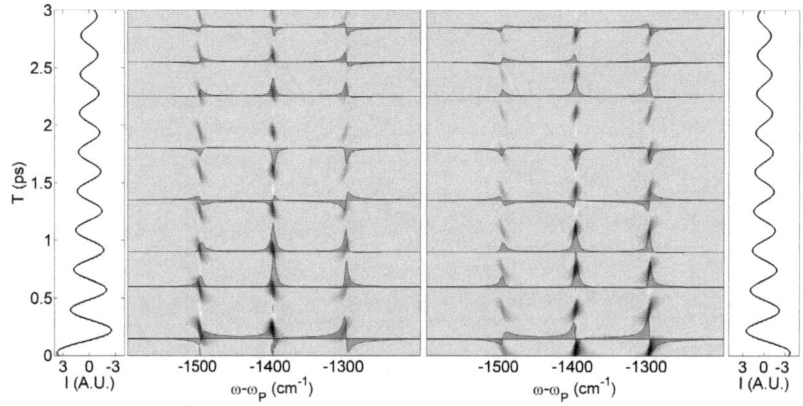

Figure 2: Two dimensional signal $S(\omega, T)$ of diagrams $d_i - d_iv$ for a system with two modes of frequency $\omega_L = 100\,\mathrm{cm}^{-1}$ and $\omega_1 = 1400\,\mathrm{cm}^{-1}$ in the red and the blue sides of the spectrum. The shaded area, which represent the one dimensional signal for selected times, emphasise the oscillation of the signal with T from loss to gain features. The side panels report the integration of the signal over the detected frequency that gives the mean number of photon.

The outcome of the simulations is shown in Fig. 2. The interplay of these features is responsible for a redistribution of photons between the red and the blue sides of the spectrum, as stressed by the side panels in Fig. 2. The total number of photons is conserved but, since red and blue components bring different energies, the total energy flows between fields and matter [14].
Then we consider a two level system, ω_1 and ω_2, with the higher close to the frequency difference between the Raman and the actinic pulse, $\omega_2 = \omega_P - \omega_A$, being ω_P and ω_A the central frequency of Raman and actinic pulse respectively. In this case, diagrams d_1-d_6 contribute for values T comparable to the actinic duration, generating dispersive lineshapes near T=0 and baseline oscillations for negative delays [15]. We show these contributions in Fig. 3.

5. Conclusions

We have introduced a general theoretical description of non resonant impulsive FSRS. In the aim of a full quantum treatment, the relevant perturbative terms contributing to the process have been isolated, tacking advantage of the loop diagram formalism, and the total fifth order

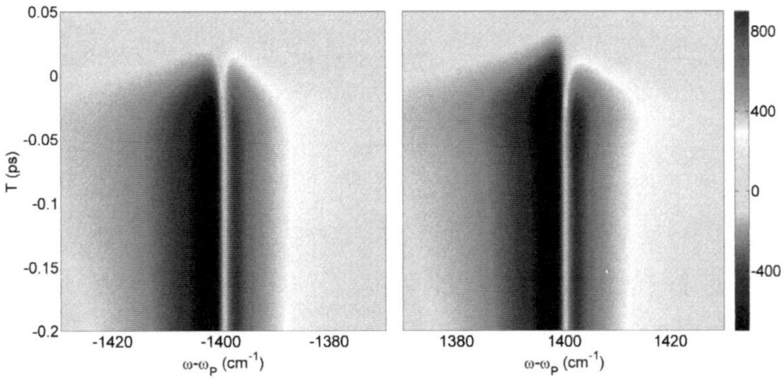

Figure 3: Two dimensional signal $S(\omega, T)$ of diagrams $d_1 - d_6$ for a system with two modes of frequency $\omega_1 = 1400\,\mathrm{cm}^{-1}$ and $\omega_2 = 5030\,\mathrm{cm}^{-1}$ in the red and the blue sides of the spectrum.

response has been calculated. In particular, we investigated the coupling of low and high-frequency modes induced by impulsive actinic excitations and we demonstrated the microscopic mechanism underlying the energy flow between spectral components and matter.
Specifically, the low-frequecy coherences created by the actinic pulse modulate the Raman signals, producing features evolving with the delay T between the actinic and probe pulses. The interplay of these features is responsible for a redistribution of photons between the red and the blue sides of the broadband pulse with respect to the narrowband Raman pulse. Although the total number of photons is conserved during the process, electromagnetic energy is exchanged between the fields and transmitted to the matter degrees of freedom.
Furthermore, we focused on the case of extremely small delays T. When the delay is comparable to the pulses duration, additional signal is generated in the pulse overlap region. This leads to unwanted nonlinear effects which can obscure sample dynamics and have to be taken under control. Indeed, we have found that the optical response in the overlap region of pump and probe fields in nr-FSRS spectroscopy exhibits time dependent features even in the absence of sample dynamics due to the absorption of the pump pulse. The correct evaluation of these contributions is a critical issue for assessing the resolution limit of pump-probe techniques and for future analysis and design of spectroscopic experiments on ultrashort time scales.

References

[1] Provencher F, Bérubé N, Parker A W, Greetham G M, Towrie M, Hellmann C, Côté M, Stingelin N, Silva C and Hayes S C 2014 **5** 4288
[2] Batignani G, Bossini D, Palo N D, Ferrante C, Pontecorvo E, Cerullo G, Kimel A and Scopigno T 2015 *Nature Photonics* **9** 506–10
[3] Kukura P, McCamant D W, Yoon S, Wandschneider D B and Mathies R A 2005 *Science* **310** 1006–9
[4] Fang C, Frontiera R R, Tran R and Mathies R A 2009 *Nature* **462** 200–4
[5] Mukamel S 1995 *Principles of Nonlinear Optical Spectroscopy* (New York: Oxford University Press)
[6] Dorfman K E, Fingerhut B P and Mukamel S 2013 *J. Chem. Phys.* **139** 124113
[7] Rahav S and Mukamel S 2010 *Adv. At., Mol., Opt. Phys.* **59** 223–63
[8] Sun Z, Fu B, Zhang D H and Lee S Y 2009 *The Journal of Chemical Physics* **130** 044312
[9] Kukura P, McCamant D W and Mathies R A 2007 *Annu. Rev. Phys. Chem.* **58** 461–88
[10] Mehlenbacher R D, Lyons B, Wilson K C, Du Y and McCamant D W 2009 *The Journal of Chemical Physics* **131** 244512
[11] Harbola U and Mukamel S 2008 *Phys. Rep.* **465** 191–222
[12] Mukamel S 2003 *Phys. Rev. E* **68** 021111
[13] Biggs J D, Voll J A and Mukamel S 2012 *Phil. Trans. R. Soc. A* **370** 3709–27
[14] Batignani G, Fumero G, Mukamel S and Scopigno T 2015 *Phys. Chem. Chem. Phys.* **17** 10454–61
[15] Fumero G, Batignani G, Dorfman K E, Mukamel S and Scopigno T 2015 *ChemPhysChem* **16** 3438–43

Transparent and conductive Al/F and In co-doped ZnO thin films deposited by spray pyrolysis

A Hadri[1], M Taibi[2], A El hat[1] and A Mzerd[1]

[1]University Mohammed V- Rabat, Faculty of Sciences, Physics Department, LPM, B.P. 1014 Agdal-Rabat, Morocco.
[2]University of Mohammed V, LPCMN, Ecole Normale Superieure Rabat-Morocco

E-mail: mzerd@yahoo.fr

Abstract. In doped ZnO (IZO), In-Al co-doped ZnO (IAZO) and In-F co-doped ZnO (IFZO) were deposited on glass substrates at 350 °C by spray pyrolysis technique. The structural, optical and electrical properties of as-deposited thin films were investigated and compared. A polycrystalline and (002) oriented wurtzite crystal structure was confirmed by X-ray patterns for all films; and the full width at half –maximum (FWHM) of (002) diffraction peak increased after co-doping. The investigation of the optical properties was performed using Uv-vis spectroscopy. The average transmittances of all the films were between 70 and 85%. Hall Effect measurements showed that the electrical conductivity of co-doped films increased as compared with IZO thin film. The highest conductivity of about 16.39 Ω^{-1} cm^{-1} was obtained for as-deposited IFZO thin film. In addition, the thin films were annealed at 350 °C for two hour under Ar atmosphere and their optical, electrical properties and the associated photoluminescence (PL) responses of selected films were analysed. After annealing, the electrical conductivity of all thin films was improved and the optical transmittance remained above 70%. Room temperature PL revealed that the annealed IAZO thin film had a strong green emission than that of IZO film.

1. Introduction

Transparent conducting oxides (TCOs) have attracted much attention owing to their capacity of transporting electric charge and transmitting photon in visible range. Research efforts were focused on ZnO-based transparent conducting oxides because of its low cost, high chemical stability and excellent optoelectronic properties [1]. These properties make ZnO suitable to be used as active materials in electronics and photonics [2,3]. The deposition of ZnO thin films have been reported using several techniques such as radio frequency magnetron sputtering [4], pulsed laser deposition [5], chemical vapor deposition [6], sol-gel [7] and spray pyrolysis [8,9]. Among these, spray pyrolysis has some advantages due to its easy scalability and inexpensive experimental arrangement. It is worth mentioning that the electrical properties of undoped ZnO are not enough to meet the increasing needs on fabricating high performance semiconductor devices [10]. Therefore, doping with group III elements such as In, Al and Ga presents an interesting alternative to improve their optoelectrical properties [11-13]. In addition, F doped ZnO thin films were reported to enhance the electrical conductivity by improving the mobility [14]. Thus, it is expected that co-doping with both In and Al/F element could lead to an improvement of different properties.

In this work, the co-doping effects of Al/F and In on ZnO thin films were firstly studied. Then, the effect of annealing on the optical and electrical properties was investigated. For comparative purposes, In doped ZnO thin films were prepared and their properties were also determined.

2. Experimental details

In doped ZnO (IZO), In-Al co-doped ZnO (IAZO) and In-F co-doping (IFZO) thin films were deposited on glass substrates using chemical spray technique. An homogenous solution was prepared by dissolving zinc chloride ($ZnCl_2$) [0.05 M] as zinc precursor and indium chloride ($InCl_3$), aluminum nitrate ($Al(NO_3)_2$), ammonium fluoride (NH_4F) as doping sources in distilled water (200ml) at room temperature. The concentration of In dopant in IZO thin film was fixed at 3 at. % and the atomic ratio of the dopant elements in IAZO and IFZO thin films were (In^{3+}=3 at. %, Al^{3+}=1 at. %) and (In^{3+}=3 at.%, F^-=1 at.%), respectively. The prepared solution was sprayed onto the heated substrates at a constant temperature of 350 °C. After deposition, the samples were annealed at 350 °C. Details of experimental procedure are published elsewhere [15].

The structural properties were studied by X-ray diffraction (XRD) with Cu Kα-radiation and scanning in a 2θ range from 20° to 60° by employing a Siemens D500. The optical transmittance measurements of the films were carried out in photon wavelength range of 300-800 nm using UV-Visible spectrophotometer (Perkilmer Lambda 900) and taking into account the glass in the reference beam. The electrical properties were determined at room temperature from Hall Effect equipment (ECOPIA HMS 3000) using van der Pauw's method [16]. The photoluminescence spectra were recorded from 350 nm to 800 nm at room temperature by a 355 nm excitation line of a frequency-tripled neodymium-doped yttrium aluminum garnet (Nd-YAG) laser.

3. Results and discussion

3.1. Structural properties

The XRD patterns of IZO, IAZO and IFZO thin films were shown in figure 1. The results showed that all deposited films were polycrystalline with a structure that fit well with the ZnO wurtzite type (JCPDS no. 89-1397). In addition, no secondary phases were observed which excludes the existence of dopant based clusters. The peaks intensities were different and depended on the nature of doping element. IZO and IFZO thin films showed a high intensity of the (002) diffraction peak compared with the other peaks suggesting that the growth orientation was along the c-axis normal to the substrate. In addition, it was apparent that IAZO thin films exhibited a highly oriented (002) diffraction peak and no diffraction from randomly oriented grains can be observed indicating an improvement of stoichiometry [17].

Figure 2 showed the full width at half –maximum (FWHM) of (002) diffraction peak as a function of as-deposited films. Values of FWHM were slightly increased. The FWHM of IAZO thin films was slightly higher than those of IZO and IFZO thin film, indicating that IAZO thin films had the smallest grain size among the as-deposited films. This behavior might be caused by the stress formation as a result of the relatively high difference of ionic radius between Al^{3+} and Zn^{2+} ($r_{Al}{}^{3+}$=0.053 nm and $r_{Zn}{}^{2+}$= 0.074 nm) as compared with F^- and O^{2-} ($r_F{}^-$=0.131 nm and $r_O{}^{2-}$=0.138 nm) [18,19]. This is in good agreements with other reports [20-21].

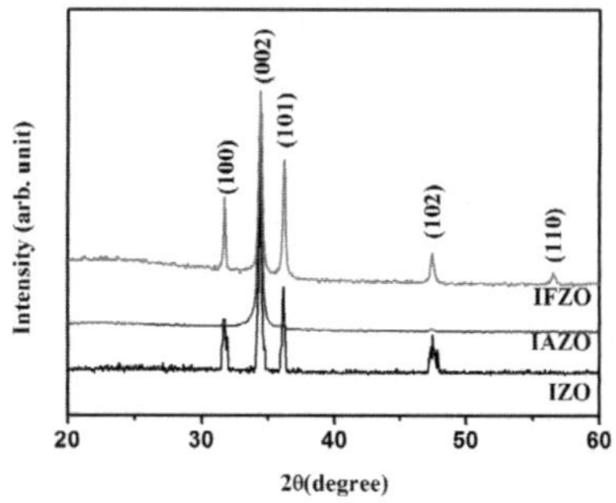

Figure 1 XRD spectra of as-deposited thin films.

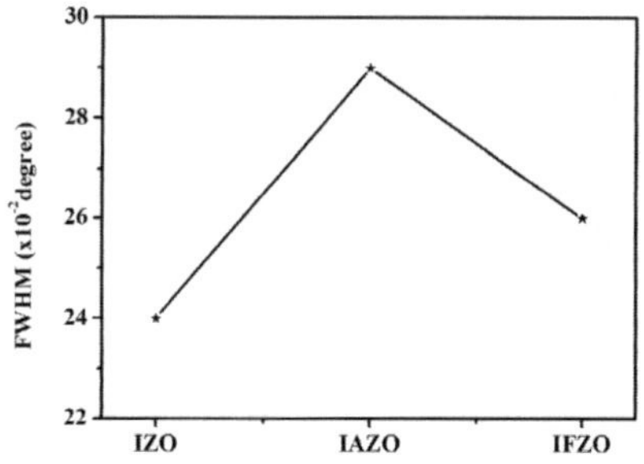

Figure 2 Variation of FWHM in the (002) direction of as-deposited thin films.

3.2. Optical properties

In order to know the effect of co-doping and annealing on the optical properties of deposited thin films, the optical transmittance was measured and shown in Figure 3. All as-deposited films showed a high transmittance between 70 % and 85 % in the visible region and exhibited a sharp absorption edges in the wavelength region at about 375 nm.

After annealing, the optical transmittance in the visible region of IZO and IAZO decreased slightly by 4%. As indicated in the electrical properties section, annealing led to a significant increase in the carrier concentration for IZO and IAZO thin films which should be responsible for the observed decrease in these films. Similar observation was reported by You et al. who attributed this to the increase of free carrier absorption [22].On the other hand, IFZO thin films exhibited a slight increase in transparency which might be due to an improvement of film crystallinity. Additionally, the annealed films resulted in a slight blue shift of the absorption edge. The observed variation could be explained by Burstein-Moss effect [23].

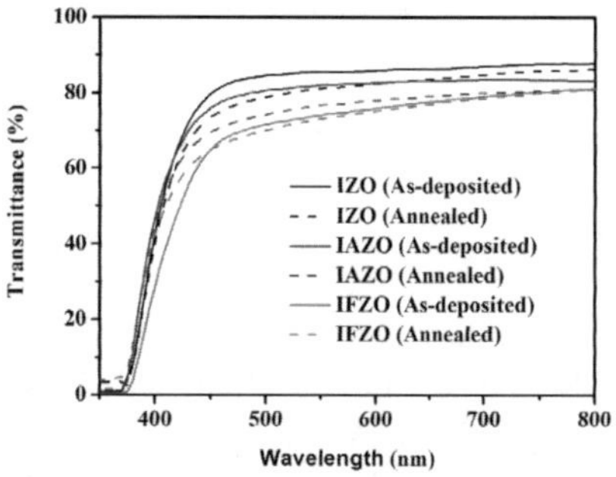

Figure 3 Optical transmittance of as-deposited and annealed thin films.

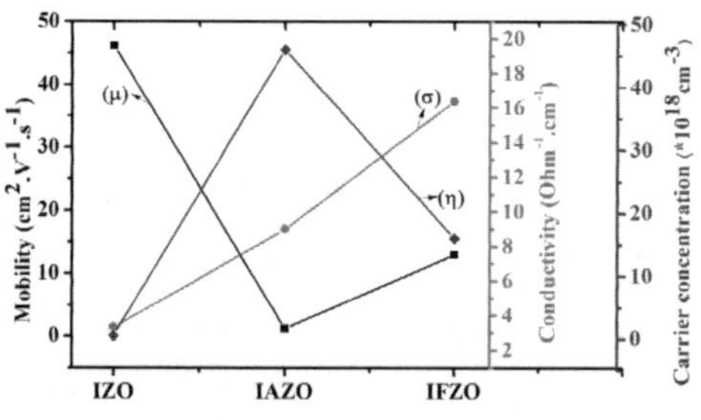

Figure 4 The electrical conductivity (σ), carrier concentration (η) and mobility (μ) of as-deposited thin films.

3.3. Electrical properties

Figure 4 showed the variation of the electrical parameters as a function of as-deposited samples. As it could be seen, the carrier concentration of both IAZO and IFZO thin films increased as compared with IZO thin films. Such behavior could be as a result of the substitution of F^- at O^{2-} site or Al^{3+}at Zn^{2+}site which provided one free electron. It is clear that the electrical conductivity of both IAZO and IFZO thin films increased. The IFZO thin films showed the highest electrical conductivity due to its high mobility as compared with IAZO thin film, which might be due to the perturbation of the valence band by fluorine ions and therefore leaves the conduction bands free of scattering [24]. In contrast, the lowest mobility was obtained for the as-deposited IAZO thin films. Since the carrier concentration of this thin film is in the range between 10^{19}-10^{20} cm^{-3}, the obtained mobility was mainly dominated by the ionized impurity scattering and grain boundary scattering [25].

The variation of conductivity and carrier concentration as a function of as-deposited and annealed thin films is shown in figure 5. The results showed that annealing led to further improvement of conductivity. The observed increase in conductivity of all thin films could be attributed to desorption of oxygen at grain boundaries, which was chemisorbed during the deposition in an oxygen containing atmosphere [26]. It worth mentioning, that the annealed IAZO thin film presented a high conductivity as compared with the annealed IFZO thin film, which could be due to the escape of volatile F from the ZnO lattice [27]. On the other hand, the most pronounced improvement in conductivity was observed for IZO thin films. The observed difference between the conductivity values of annealed IZO and IAZO films could be explained by the rearrangement of aluminum and indium ions occupying substitutional position instead of interstitial. Meanwhile, the radius of In^{3+} (0.08 nm) is closer to Zn^{2+} (0.074 nm) than that of Al^{3+}(0.053 nm). Therefore, the substitution of Zn^{2+} by Al^{3+} might cause large lattice deformation which could be responsible of generating more crystal defects.

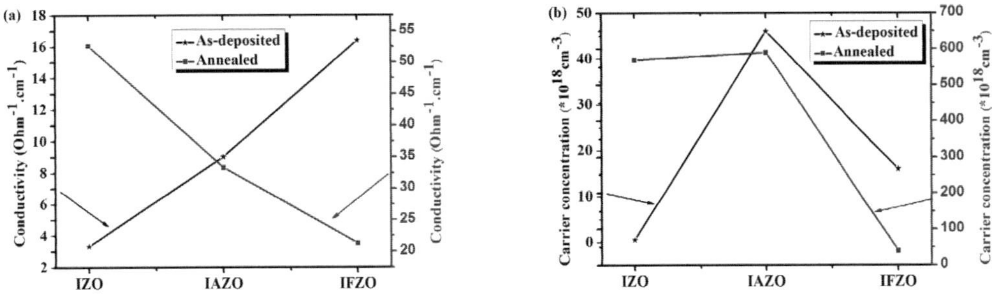

Figure 5 Variation of the electrical properties of as-deposited and annealed thin films as a function of: (a) Conductivity (b) carrier concentration.

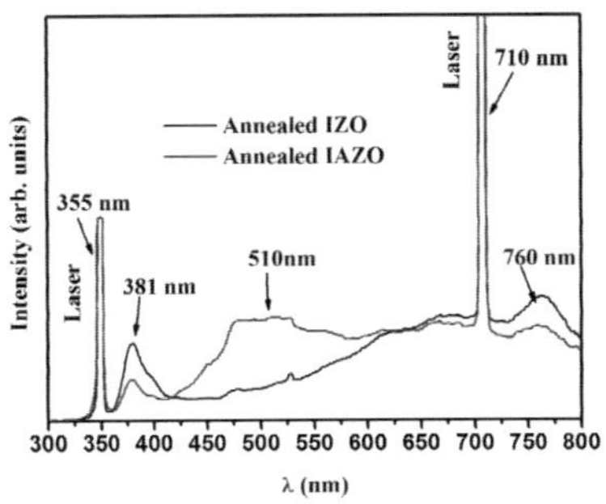

Figure 6 Photoluminescence spectra of annealed IZO and IAZO Thin films.

3.4. Photoluminescence studies

PL study is an important method to investigate the levels of defect by analyzing the radiative recombination of the samples. Taking into account the electrical behavior of our films, IZO and IAZO annealed thin films were chosen to study their photoluminescence properties and are presented in figure 6. The intense peak at 355 nm and its second order at 710 nm were attributed to the laser source, while the peak at 381 nm corresponds to excitonic recombination corresponding to the near-band edge (NBE) emission of ZnO [28]. The second order of this peak is observed in 760 nm. A broad peak centered at about 510 nm was associated to deep level emission (DLE) and was related to the crystallinity and the lattice structural defects such as oxygen vacancies and dislocation [29]. The annealed IAZO thin film showed a strong broad band of DLE which implied that the concentration of defects was very high. In contrast, the annealed IZO thin film almost passivated the oxygen based point defect and increased the NBE, indicating a good crystallinity of this film. This behavior strengthened the assumption used in the electrical properties concerning the effect of ionic radius difference.

Conclusion

In this work, IZO, IAZO and IFZO thin films were deposited onto glass substrates by spray pyrolysis technique, and the structural, optical and electrical properties of the samples were compared and discussed. In addition, the effect of annealing at 350 °C under Ar atmosphere on the optical and electrical properties was also investigated. All as-deposited films were polycrystalline with a (002) preferential orientation. The optical transmittance of all samples was above 70 % in the visible region. The as-deposited IFZO thin film had the highest electrical conductivity which was about 16.4 $\Omega^{-1}cm^{-1}$. After annealing, the IAZO thin film had better conductivity than the IFZO thin film, which was around 33.3 $\Omega^{-1}cm^{-1}$ combined with an optical transmittance above 75%. Moreover, the annealed IZO thin film exhibited the highest electrical conductivity and optical transmittance. The PL results showed that the optical qualities of IZO thin film were better than IAZO thin film, which was attributed to the lattice deformation introduced by larger ionic radius differences.

References

[1] Djurišić A B, Ng A M C and Chen X Y 2010 *Prog. Quant. Electron.* **34** 191–259
[2] Ryu R Y, Lee T S, Lubguban J A, White H W, Kim B J and Park Y S 2006 *Appl. Phys. Lett.* **88** 241108
[3] Fan J C, Sreekanth K M, Xie Z, Chang S L and Rao K V 2013 *Prog. Mater. Sci.* **58** 874–985
[4] Purohit A, Chander S, Sharma A, Nehra S P and Dhaka M S 2015 *Opt. Mater.* **49** 51–8
[5] Sahu R, Dileep K, Negi D S, Nagaraja K K, Shetty S and Datta R 2015 *J. Cryst. Growth* **410** 69–76
[6] Wang X H, Huang L Q, Niu L J, Li R B, Fan D H, Zhang F B, Chen Z W, Wang X and Guo Q X 2015 *J. Alloy. Compd.* **622** 440–5
[7] Cittadini M, Sturaro M, Guglielmi M, Resmini A, Tredici I G, Anselmi-Tamburini U, Koshy P, Sorrell C C and Martucci A 2015 *Sensors Actuat. B* **213** 493–500
[8] Hadri A, Loghmarti M, Mzerd A and Taibi M 2014 *International Renewable and Sustainable Energy Conference* (Ouarzazate) p 61-5
[9] Hadri A, Nassiri C, Chafi F Z, Loghmarti M and Mzerd A 2015 *Energy and Environment Focus* **4** 12-7
[10] Yang Y, Qi J J, Liao Q L, Zhang Y, Yan X Q, Huang Y H and Tang L D 2009 *Appl. Phys. A* **94** 799-803
[11] Hadri A, Nassiri C, Chafi F Z, Loghmarti M, Tricha M and Mzerd A 2015 *Physical and Chemical News* **75** 1-6
[12] Lee H N, Shin C H, Hwang D K, Kim H, Oh K and Kim H J 2015 *J. Alloy. Compd.* **644** 193–8
[13] Kuprenaite S, Murauskas T, Abrutis A, Kubilius V, Saltyte Z and Plausinaitiene V 2015 *Surface and Coatings Technology* **271** 156–64
[14] Zhang X, Zhu L, Xu H, Chen L, Guo Y and Ye Z 2014 *J. Alloy. Compd.* **614** 71–4
[15] Hadri A, Nassiri C, Chafi F Z, Loghmarti M, Fares B, Laanab L, Chraibi F, Bensitel M and Mzerd A 2014 *Sensors & Transducers* **178** 63-8
[16] Matsumura T and Sato Y 2010 *Journal of* Modern *Physics* **1** 340-7
[17] Yuste M, Galindo R E, Caretti I, Torres R and Sanchez O 2012 *J. Phys. D Appl. Phys.* **45** 025303
[18] Huang C C, Wang F H and Yang C F 2013 *Appl. Phys.* A **112** 877–83
[19] Mallick A, Sarkar S, Ghosh T and Basak D 2015 *J. Alloy. Compd.* **646** 56–62
[20] Yoon H S, Lee K S, Lee T S, Cheong B, Choi D K, Kim D H and Kim W M 2008 *Sol. Energ.*

Mat. Sol. C. **92** 1366–72
[21] Sengupta J, Sahoo R K and Mukherjee C D 2012 *Mater. Lett.* **83** 84–7
[22] You Q, Cai H, Gao K, Hu Z, Guo S, Liang P, Sun J, Xu N and Wu J 2015 *J. Alloy. Compd* **626** 415–20
[23] Kim Y, Lee W, Jung D R, Kim J, Nam S, Kim H and Park B 2010 *Appl. Phys. Lett.* **96** 171902
[24] Gordon R G 2000 *MRS Bull.* **25** 52–7
[25] Minami T, Suzuki S and Miyata T 2001 *Thin Solid Films* **398–399** 53-8
[26] Orton J W and Powell M J 1980 *Rep. Prog. Phys.* **43** 1263
[27] Seo J S, Jeon J H, Hwang Y H, Park H, Ryu M, Park S H K and Bae B S 2013 *Sci. Rep.* **3** 1–9
[28] Ameen S, Akhtar M S, Seo H K, Kim Y S and Shin H S 2012 *Chem. Eng. J.* **187** 351–6
[29] Shewale P S, Sim K U , Kim Y B, Kim J H, Moholkar A V and Uplane M D 2013 *J. Lumin.* **139** 113–8

Journal of Physics: Conference Series **689** (2016) 012025 doi:10.1088/1742-6596/689/1/012025

Electronic properties of $Hg_{1-x}Cd_xSe$ lens-shaped quantum dots under external fields

J R Herrera[1], W Gutierrez[1,2], D A Miranda[1]

[1]CIMBIOS, Escuela de Física, Facultad de Ciencias, Universidad Industrial de Santander, Cr. 27 Cll. 9, Bucaramanga, Colombia

[2]FICOMACO, Escuela de Física, Facultad de Ciencias, Universidad Industrial de Santander, Cr. 27 Cll. 9, Bucaramanga, Colombia

E-mail: joseherreraibagos@gmail.com, dalemir@uis.edu.co

Abstract. $Hg_{1-x}Cd_xSe$ are II-VI semiconductors alloys with optoelectronic properties that depend upon the molar fraction x, which can be further controlled by nanostructuring. In this work one electron confined in a zero-dimensional lens-shaped nanostructure of $Hg_{1-x}Cd_xSe$ surrounded by a matrix of different molar fraction is analyzed and its electronic properties are studied under external magnetic and electric fields. Our system was modeled by means of the 3D Schrödinger equation in the framework of the effective mass approximation, which was solved using a finite element method. The model is described by a discontinuous space with Ben Daniel-Duke boundary conditions. We calculated the energy spectrum and the corresponding probability density of the electron for some low-lying energy levels as a function of: electric field strength on plane and magnetic field strength applied along the growth direction. Also, the effect of finite confinement potential was studied in presence of a uniform magnetic field. Our results shown that the electronic properties of $Hg_{1-x}Cd_xSe$ quantum dots are highly sensitive to a threading magnetic field because the degenerate energy levels are split. On the other hand, the effect of electric and magnetic fields applied simultaneously on a quantum dot can increase the system stability against external perturbation, e.g. thermal interactions.

1. Introduction

The recent interest in semiconductor nanostructured systems has grown side by side with the expansion of technological boundaries and the search of small, light and economic electronic devices, which are used on a wide spectrum of applications. This has lead to the study of material that show interesting opto-electronic properties, such as $Hg_{1-x}Cd_xSe$ alloys. They detect electromagnetic radiation from the infrared to the visible spectrum through molar fraction variations. Also, this alloy can be fabricated with fewer defects than the more known and studied $Hg_{1-x}Cd_xTe$ alloys [1].

The study of this alloy started in the 70s [2] and has attracted the attention of military contractors for the study and future development of infrared detection devices [1]. Recent experimental and theoretical studies on the $Hg_{1-x}Cd_xSe$ alloy showed its potential in the development of new opto-electronic technologies [3–8].

The electronic band structure of $Hg_{1-x}Cd_xSe$ alloys was calculated by [9], and it shows a strong dependence on the molar fraction. When $x = 0$ it behaves like a semi-metal, in $x \approx 0.1$ it suffers an inversion between the conduction and valence bands, and for $x \gtrsim 0.1$ is a semi-conductor.

The effective mass of the alloy was given by [2], where it is shown that its effective mass is given by a relation with the energy gap. This relation makes it interesting to analyze the semi-conductor alloy with values close to $x \approx 0.1$, in this case the effective mass has a very small value, which translates in a strong quantum behavior.

2. Theoretical basis

Due to the huge amount of atoms which compose a Quantum Dot (QD), it will be difficult to calculate the interaction of the electrons with each one of the atoms that compose the lattice. We considered a single interaction of the band structure between the valence and conduction band, the effective mass approximation. Under this approximation the Hamiltonian describes the quantum behavior of an electron confined inside a tridimensional QD under the influence of magnetic and electric fields

$$\hat{H} = \frac{1}{2m^*}\left[\vec{p}(x,y,z) - e\vec{A}(x,y,z)\right]^2 + eFx + V(x,y,z). \tag{1}$$

The Coulomb gauge $\vec{A} = \frac{1}{2}\left(\vec{B}\times\vec{r}\right)$ was used to express the magnetic potential $\vec{A}$ in terms of the magnetic field. This gauge is essential to properly describe the charge interaction with the magnetic field and is used in semiclassical calculations. The possible energy levels of the system are the eigenvalues of the Hamiltonian. Considering an oriented magnetic field on a θ angle relative to the direction of growth and an oriented electric field along the QD plane, we obtain the dimensionless equation of Schrödinger-Pauli

$$\left\{-\nabla^2 - i\gamma\left[y\left(\cos\theta\frac{\partial}{\partial z} - \sin\theta\frac{\partial}{\partial x}\right) + (x\sin\theta - z\cos\theta)\frac{\partial}{\partial y}\right] + \frac{\gamma^2\left[y^2 + (x\sin\theta - z\cos\theta)^2\right]}{4} + \alpha Fx + V(x,y,z)\right\}\psi_n(x,y,z) = E_n\psi_n(x,y,z). \tag{2}$$

Where $\gamma = \frac{e\hbar B}{2m^* R_y^*}$ is the dimensionless coefficient of the magnetic field, that is present in the second and third term as the paramagnetic and diamagnetic parts, $\alpha = \frac{ea_0^*}{R_y^*}$, the dimensionless coefficient of the electric field, $a_0^* = \frac{\hbar\epsilon}{m^* e^2}$, the effective Böhr radius, $R_y^* = \frac{e^2}{2\epsilon a_0^*}$, the effective Rydberg and $V(x,y,z)$ represents the structural confinement potential

3. Modeling and simulation by finite elements

Two models were used to perform calculations. In the first model, the QD was considered isolated, with infinite confinement potential outside the QD and zero inside the QD, as can be seen in Fig. 1.

$$V(x,y,z) = \begin{cases} 0 & \text{Inside the QD} \\ \infty & \text{Outside the QD} \end{cases} \tag{3}$$

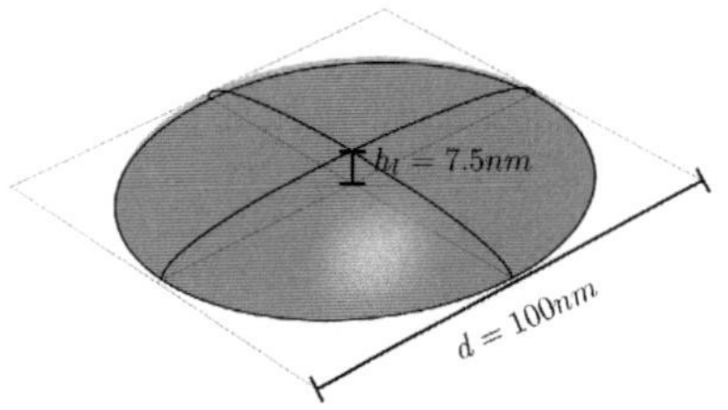

Figure 1: Lens-shaped QD.

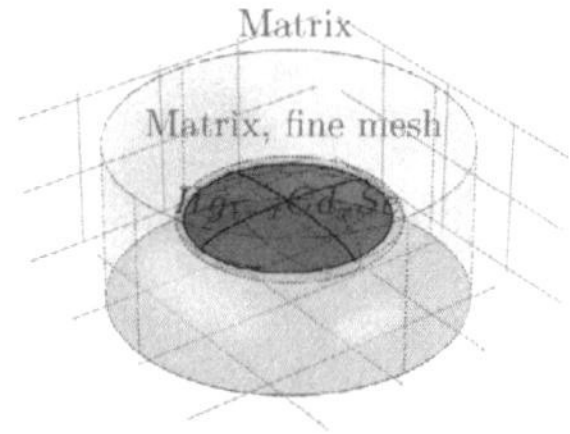

Figure 2: QD surrounded by a $Hg_{1-x_0}Cd_{x_0}Se$ matrix.

In the second model a finite confinement potential was considered. In this model the nanostructure is surrounded by an alloy with different molar fraction. Two zones are defined, the first one representing the QD of $Hg_{1-x}Cd_xSe$, molar fraction $x = 0.14$, and the second one representing the matrix that covers the QD, with a different molar fraction $Hg_{1-x_0}Cd_{x_0}Se$, $x_0 = 0.18$, this zone is simulated by a cylinder.

The problem was solved in a discontinuous space meshed of tetrahedral finite elements, the shape function used to interpolate over the space is a Lagrange polynomial of third degree. The meshing parameters were calculated by a convergence criteria which assures three significant figures; their values were chosen based on [10].

Due to the fact that the confinement potential is finite there is a possibility of finding the electron outside the QD; a very fine mesh was used in the matrix zone that is closer to the QD. This helps in the optimization of calculation time based on the criteria that are more important the QD domain and its surroundings. A representation of this can be seen in Fig. 2.

The confinement potential is given by Eq. 4, the potential in the matrix is calculated as the 60% of the difference between the energy gap of the QD and the energy gap of the matrix.

$$V(x,y,z) = \begin{cases} 0 & \text{Inside the QD} \\ V = 39\,\text{meV} & \text{Inside the matrix} \\ \infty & \text{Otherwise} \end{cases} \tag{4}$$

The effective mass of the electron inside the QD has a value m^*_{QD}, on the matrix its value is given by m^*_X.

$$m^* = \begin{cases} m^*_{QD} = 0.08m_e & \text{Inside the QD} \\ m^*_X = 0.51m_e & \text{Inside the matrix} \end{cases} \tag{5}$$

The wave function that describes the electron needs to be continuous over the space, however, the properties that describe the behavior of this electron are different in each alloy. An appropriate boundary condition is the Ben Daniel-Duke condition [11], which is used to ensure continuity in the interface between the QD and the matrix. The following equation takes into account the change of effective masses on this boundary

$$\frac{1}{m^*_{QD}}\nabla\psi(x,y,z)\cdot\vec{n} = \frac{1}{m^*_X}\nabla\psi(x,y,z)\cdot\vec{n}. \tag{6}$$

On the boundaries, when the confinement potential is infinite, the wave function takes null values; in such points the Dirichlet boundary condition is imposed

$$\psi = 0. \tag{7}$$

Journal of Physics: Conference Series **689** (2016) 012025 doi:10.1088/1742-6596/689/1/012025

4. Numerical results and discussions

In order to study the effect due to the application of external magnetic and electric fields on the electronic properties of a confined electron inside a QD, the energy levels calculations were performed to the first and second models described in the previous section. The calculation includes energy spectrum as a function of the magnetic field intensity over the growing direction and electric field along the plane of the QD, as a function of the angular orientation of the magnetic field between the growing direction and along QD plane. The probability density for the lowest states was calculated to 3D, however a 2D representation were obtained by the integration along the growing axis.

4.1. First model: infinite potential

The lowest energy levels of one-electron QD as a function of the magnetic field intensity are shown in Fig. 3. The observed behavior in the energy levels shows several features: crossing points between different levels, energy levels monotonically growing, initially increase and then decrease of energy levels. Such behavior can be understood as the superposition of the two terms due to the interaction of the electron with the magnetic field. The paramagnetic term defines the coupling between the magnetic field and the permanent magnetic moment of the rotating electron that is linear and can be positive or negative depending on if its clockwise or counter-clockwise direction. The diamagnetic term is always positive, quadratic and increases the kinetic energy of the electron.

The lens axial symmetry promotes degeneration in the energy system, which is evident in the number of crosses between energy levels, in the low field intensity regime the structural and the magnetical components of the confinement are comparable. It is also observed how the levels that were degenerate in the absence of magnetic field are now unfolded under the effect of an external field, analogous to the Zeeman effect which is observed in atoms, giving more accessible states to the electrons. Furthermore, as the magnetic field increases the spectral structure acquires a Landau levels type configuration associated to a dominant magnetic confinement.

The energy levels in absence of external field are similar to the quantum harmonic oscillator. This is explained by the similarity in the shape of the confinement potential that acts over the electron. In the Fig. 3 is shown how the upper energy levels lose their degeneracy.

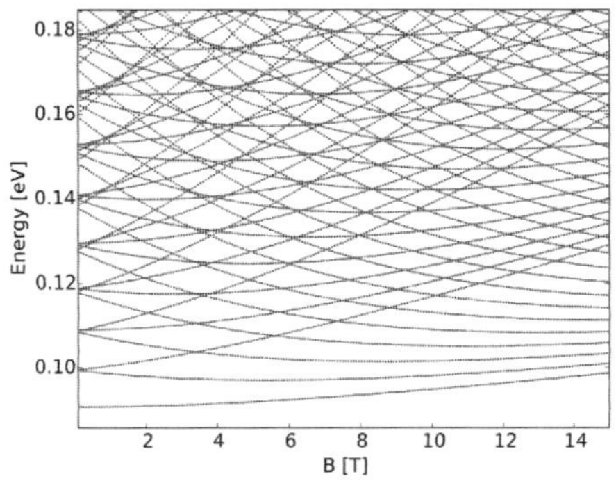

Figure 3: Energy levels for a lens-shaped QD under external magnetic field B on the growing direction with infinite potential confinement.

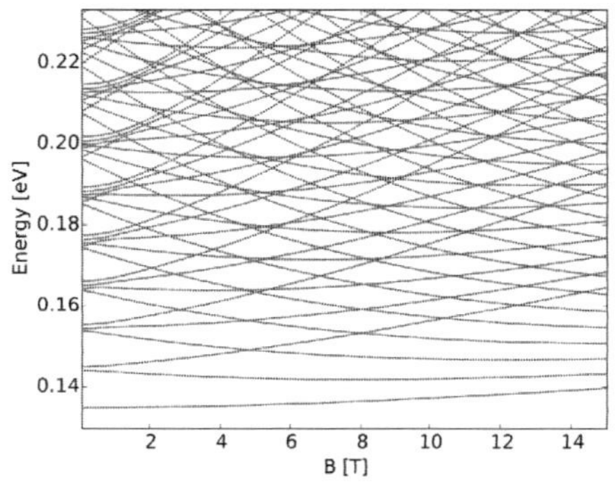

Figure 4: Energy levels for a lens-shaped QD under external magnetic field B on the growing direction and electric field on the plane direction with infinite potential confinement.

The lowest energy levels for a QD as a function of the magnetic field intensity with constant electric field are shown in the Fig. 4. The lens has a low height, an additional confinement in the direction of growth does not generate changes in energy spectra; so the field direction was simulated in the plane of the QD. An additional confinement due to the electric field, which greatly increases the energy of all levels, is observed. In the other hand, the states are unfolded in absence or for weak magnetic fields because the electrical confinement breaks the rotational symmetry of the lens. When the magnetic field contribution to the energy is greater than the electrical contribution, the energy levels behave in a similar way to what is seen in Fig. 3.

4.2. Second model: finite potential

In Fig. 5 the lowest energy levels for the lens-shaped model with finite barriers are shown. A limit on the number of located states within the structure is appreciable, and is given by the confinement potential between the two alloys and the magnetic confinement. Higher energy levels than this confinement extend the probability of finding the electron over the matrix and behave quasi-continuously because in the simulation this matrix has a finite dimension. If the nanostructure was embedded in an infinite crystal no quantum effects would be observed for the electron and the energy levels would be continuous.

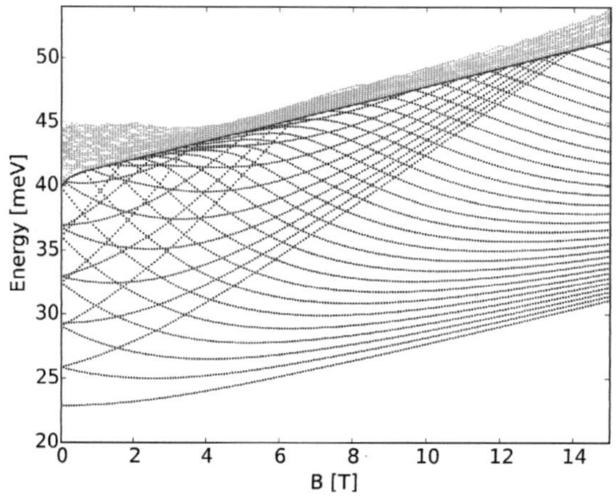

Figure 5: Energy levels for a lens-shaped QD under an external magnetic field B on the growing direction with finite potential confinement.

In Fig. 6 the probability density in the xy plane, in the xz plane and the probability percentage of finding the electron inside the QD for the first five energy levels is shown. The results reveal that the electron has a significant probability of being found outside the nanostructure; even if its energy is less than to the potential barrier, this behavior shows that the confined electron is tunneling. For large magnetic fields confinement increases towards the center of the QD, which translates in a slight increase in the probability of finding it in the structure.

It is useful to clarify that the molar fractions were selected to allow the confined electron to have an appreciable tunneling to the matrix and a limit on the number of possible states, this happens in a confining potential of only a few meV. When calculating the confinement potential equation in a molar fraction of $x = 1$, i. e. $CdSe$, a confinement potential of approximately 1eV is obtained; with this kind of confinement results are very similar to those found with an infinite potential barrier, except for a slight decrease in energy levels.

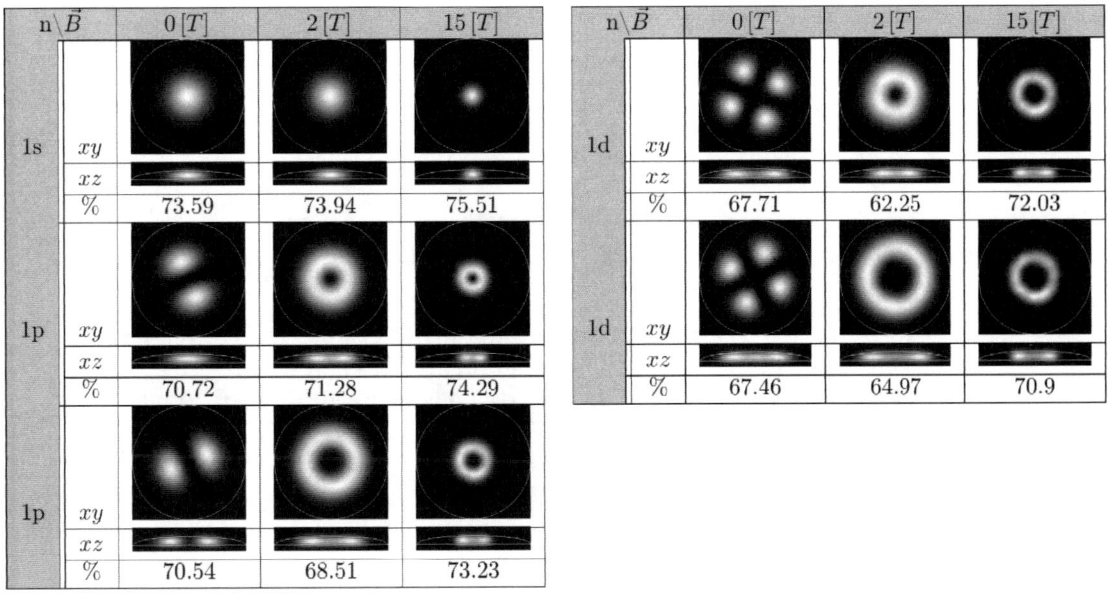

Figure 6: Probability distribution on the planes xy, xz and the probability of finding the electron inside the QD.

5. Conclusions

The results obtained for the case of an electron confined in a zero-dimensional nanostructure $Hg_{1-x}Cd_xSe$, show that the electronic properties of such systems are highly sensitive to the intensity of external fields applied. Another feature that might have great applicability is simultaneous action of magnetic and electric fields on the lens. This is of great importance because the increased energy at all levels ensures a greater system stability against external perturbation. Finally, the simulation with finite potential confinement showed tunneling effects and a limited number of states in the QD, which is what is expected for a real nanostructure.

[1] Brill G, Chen Y and Wijewarnasuriya P 2011 *J. Electron. Mater.* **40** 1679–1684
[2] Whitsett C, Broerman J and Summers C 1981 *Crystal Growth and Properties of $Hg_{1-x}Cd_xSe$* vol 16 (Academic Press)
[3] Miranda D 2011 *Caracterizacion de las Aleaciones Cristalinas $Hg_{1-x}Cd_xSe$, para $x \leq$* 0.5 Ph.D. thesis Universidad de los Andes
[4] Miranda D A, R L and Romero A H 2013 *AIP Conf. Proc* **1566** 95–96
[5] Singh K and Tanveer M 1993 *J. Mater. Sci.* **12** 737–738
[6] Himei N and Muto J 2000 *J. Mater. Sci. Mater. Electron.* **11** 145–149
[7] Hankare P, Bhuse V, Garadkar K, Delekar S and Bhagat P 2004 *Semicond. Sci. Technol.* **19** 277
[8] Sánchez M, González F, Miranda D and Meléndez Á 2015 *JPCS* vol 582 p 012046
[9] Nelson D A, Broerman J G, Summers C J and Whitsett C R 1978 *Phys. Rev. B, PRB*
[10] Markiewicz M and Voss H 2006 *Lect. Notes. Comput. SC* (Springer) pp 684–693
[11] Deyasi A and Bhattacharyya S 2011 *IJSCE* **1** 67–71

Journal of Physics: Conference Series **689** (2016) 012026 doi:10.1088/1742-6596/689/1/012026

Spectroscopy of low lying states in ^{136}Cs

B Rebeiro[1], S Triambak[1,2], R Lindsay[1], P Adsley[2,3], C Burbadge[4], G Ball[5], V Bildstein[4], T Faestermann[6], PE Garrett[4], R Hertenberger[7], A Radich[4], E Rand[4], A Varela[4], H-F Wirth[7]

[1] Department of Physics, University of the Western Cape, P/B X17, Bellville ZA-7535, South Africa.
[2] iThemba Laboratory for Accelerator Based Sciences, P.O. Box 722, Somerset West 7129, South Africa.
[3] Department of Physics, University of Stellenbosch, Matieland 7602, South Africa.
[4] Department of Physics, University of Guelph, Guelph, Ontario N1G 2W1, Canada.
[5] TRIUMF, 4004 Wesbrook Mall, Vancouver, British Columbia V6T 2A3, Canada.
[6] Physik Department, Technische Universität München, D-85748 Garching, Germany.
[7] Fakultät für Physik, Ludwig-Maximilians-Universität München, D-85748 Garching, Germany.

E-mail: b.rebeiro@gmail.com

Abstract. The low-lying excited states in ^{136}Cs relevant to the double beta decay of ^{136}Xe were studied via a ^{138}Ba(d, α)^{136}Cs transfer reaction with a high resolution magnetic spectrometer. Preliminary results from the experiment are presented.

1. Introduction

Within the standard model of particle physics [1] neutrinos are assumed to be massless. However, the observation of neutrino oscillations [2, 3] shows that the neutrinos have non-zero mass. One important goal in current physics research is to determine this mass. The experimental observation of neutrinoless double beta ($0\nu\beta\beta$) decay could provide relevant information in this regard. The $0\nu\beta\beta$ decay is possible only if the neutrino coincides with its own anti-particle (and therefore is a Majorana fermion) and it occurs when a Z=even, N=even parent nucleus is more bound than its daughter, so that the single beta decay is energetically forbidden. In this case, a two neutrino double beta ($2\nu\beta\beta$) decay to the 'granddaughter' nucleus would still be energetically possible. The $2\nu\beta\beta$ decay is a second-order standard-model-allowed transition that occurs with the emission of two electrons and two neutrinos. On the other hand, if the neutrino were a Majorana fermion, a double beta decay could also occur with the emission of two electrons but no neutrinos. This is the $0\nu\beta\beta$ decay, a standard-model-forbidden decay mode that has not been observed so far. Once experimentally measured, the $0\nu\beta\beta$ decay rate directly relates to the so called "Majorana mass" $\langle m_\nu \rangle$ through Fermi's golden rule,

$$\Gamma^{0\nu} = G^{0\nu}(Q,Z)|M^{0\nu}|^2 \frac{\langle m_\nu \rangle^2}{m_e^2}. \tag{1}$$

In the above equation, $M^{0\nu}$ is the nuclear matrix element (NME) for the decay, $G^{0\nu}(Q,Z)$ is the phase space factor and m_e is the electron mass. The dominant uncertainty in translating a

Journal of Physics: Conference Series **689** (2016) 012026 doi:10.1088/1742-6596/689/1/012026

limit on the $0\nu\beta\beta$ decay amplitude into the corresponding neutrino Majorana mass arises from theoretical limitations in calculating the nuclear matrix elements, which depend on the structure of the nuclei involved in the decay. When calculated using different methods, depending on the choice of NME calculation, the upper limit obtained on the Majorana mass is found to vary significantly [4]. For instance, a determination of the lower limit on the $0\nu\beta\beta$ decay half-life of ^{136}Xe translates into a Majorana neutrino mass smaller than 0.2-0.4 eV [5].

2. NME calculations for ^{136}Xe $\beta\beta$ decays

The NMEs for double beta decays are usually calculated (using perturbation theory) as a sum over a range of intermediate states $|k\rangle$ with energy E_k. In particular, in the case of $2\nu\beta\beta$ decay, the matrix elements $M^{2\nu}$ are calculated as a sum of virtual Gamow-Teller excitations through 1^+ states in the intermediate nucleus,

$$M^{2\nu} = \sum_{k,l,m} \frac{\langle 0_f^+|\sigma_l\tau_l^+|k\rangle\langle k|\sigma_m\tau_m^+|0_i^+\rangle}{E_k - (M_i - M_f)/2}, \tag{2}$$

where $M_i - M_f$ is the total energy released in the decay.
On the other hand, in the case of $0\nu\beta\beta$ decay the NME connects all the multipoles of the intermediate nucleus. In such decays the virtual neutrinos transfer a large momentum ($\sim 100 - 200$ MeV), allowing for nuclear excitations in the intermediate nuclei upto very high energies. This makes the matrix element calculations computationally expensive and challenging. The problem is bypassed in most calculations by using the completeness relation, where the energies of the intermediate states is replaced by an average energy, $\langle E\rangle$. This assumption, called the closure approximation, is extensively used for the NME calculations. It is now well known that the closure approximation introduces an uncertainty of $\sim$ 10% in the computation [6, 7]. Therefore, recent shell model calculations have used the knowledge of low-lying intermediate states in a truncated model space in order to have more precise calculations of the $0\nu\beta\beta$ matrix elements [8]. Thus, any knowledge of the low-lying states in the intermediate nuclei is important to place double beta decay matrix elements on a more secure footing.

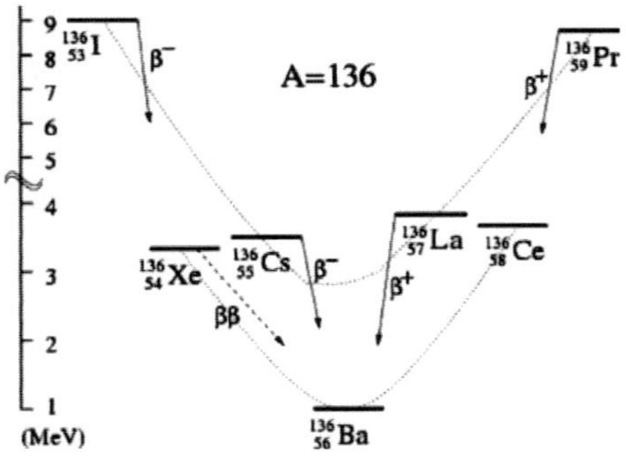

Figure 1. Mass parabola for A=136 showing the double beta decay candidates. Figure taken from Ref. [4].

The decay of ^{136}Xe to ^{136}Ba is a promising candidate for the observation of the $0\nu\beta\beta$ decay in both ongoing and proposed experiments, such as, EXO [9], Kamland-ZEN [10] and XMASS [11]. Until recently, experimental information regarding the states in the intermediate ^{136}Cs nucleus (Fig. 1) was limited to the ground state and to an isomeric 8^- state. In 2011, the experimental information on the excited 1^+ states in ^{136}Cs and the corresponding Gamow-Teller strengths (the $B(GT)$ values) relevant for the NME calculation of the ^{136}Xe $\beta\beta$ decay were obtained using the Grand Raiden spectrometer in RCNP (Japan) via the ^{136}Xe(^{3}He, t) reaction [12]. The tritons were measured on the focal plane of the spectrometer, with an energy resolution of ~ 40 keV to obtain the total Gamow-Teller strength. Since the level density in ^{136}Cs is expected to be relatively high, we report a new measurement of the excited states in ^{136}Cs with a higher resolution (~ 10 keV) using the ^{138}Ba(d, α) reaction. In our case, the reaction selectivity was not restricted only to the 1^+ excitations and other multipoles were populated as well. It is anticipated that our results will aid NME calculations for the $0\nu\beta\beta$ decay of ^{136}Xe beyond the closure approximation.

3. Experimental details

The states of interest in ^{136}Cs were produced using the ^{138}Ba(d, α) reaction with a 22 MeV, 500 nA deuteron beam from the tandem accelerator at Maier-Leibnitz-Laboratorium (MLL) in Garching (Germany). The beam was incident on a 99.9% enriched 40μg/cm^2-thick Ba_2CO_3 target evaporated on a carbon foil. The reaction products were momentum analyzed with a high resolution one quadrupole three dipole (Q3D) magnetic spectrometer [13] and detected on a 1-m long cathode-strip focal plane detector [14]. The quadrupole lens of the Q3D was used to focus and optimize the resolution of the ejectiles from the desired reaction channel while the 3 dipole magnets allowed to seperate the ejectile trajectories based on their magnetic rigidity (momentum-to-charge ratio) on the focal plane. The focal plane detector consisted of two proportional counters filled with isobutane gas at 500 mbar to measure the energy loss of the incident particles and a 7mm thick plastic scintillator to completely stop them. One of the proportional counters was coupled to a cathode-strip foil in order to determine the position of the light ejectiles with a precision better than 0.1 mm. Fig. 2 shows schematic views of the Q3D magnetic spectrometer and of the focal plane detectors. The alpha particle spectra

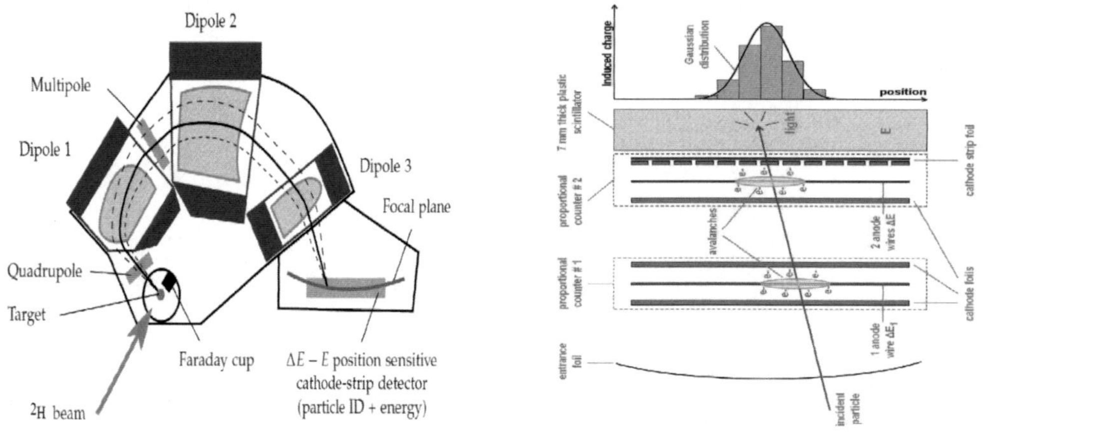

Figure 2. Left panel: The Munich Q3D spectrometer. Right panel: Schematic of the focal plane detectors.

were collected at laboratory angles ranging between 5° and 45°. We also performed angular distribution measurements for elastically scattered deuterons between an angular range of 10° and 60°. The focal plane spectrum was calibrated using well known states in ^{90}Y produced by the ^{92}Zr(d, α) reaction. Fig. 3 shows the spectra from the ^{138}Ba(d, α)^{136}Cs and ^{92}Zr(d, α)^{90}Y reactions respectively.

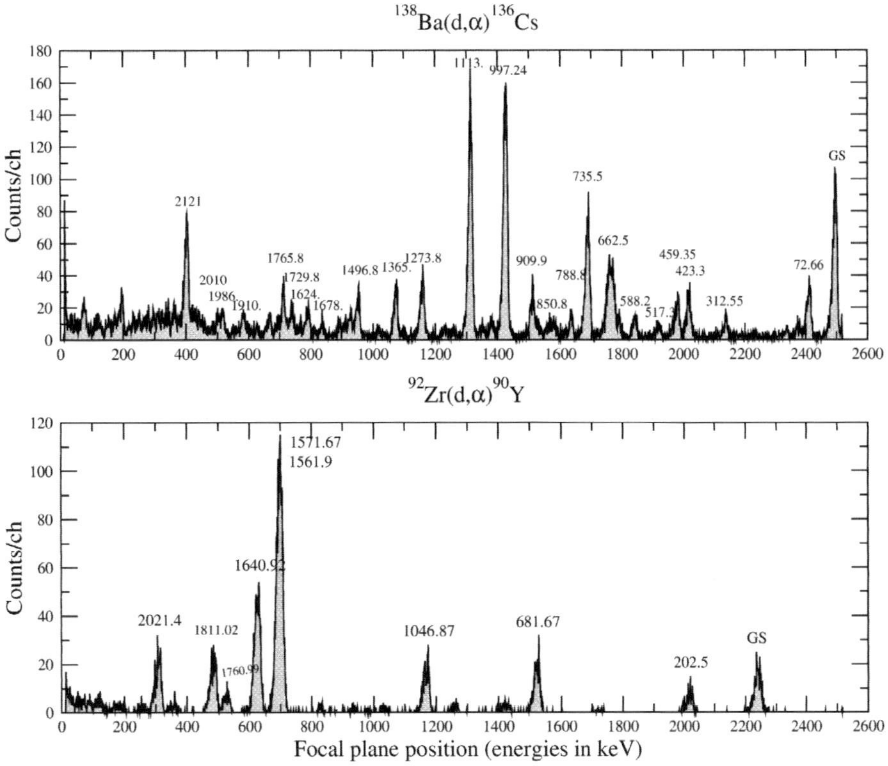

Figure 3. Preliminary results. Top panel: focal plane spectrum for ^{138}Ba(d, α)^{136}Cs. Bottom panel: ^{92}Zr(d, α)^{90}Y calibration spectrum.

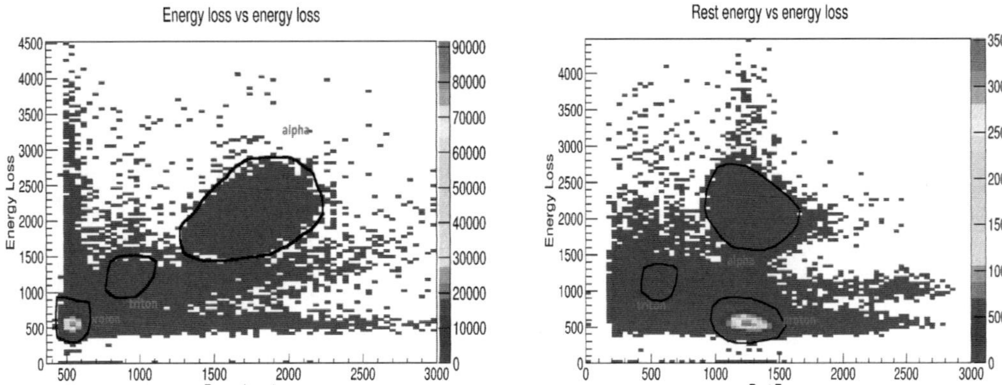

Figure 4. Particle identification spectra - left panel: $\Delta E_1 - \Delta E$, energy lost by the ejectiles in the two proportional counters and right panel: $\Delta E - E$ energy lost in the second proportional counter vs total energy deposited when ejectile loses all of its energy in the plastic scintillator. The various ejectiles produced are identified and gates are set on the alpha particles corresponding to the ^{138}Ba(d, α) reaction channel.

4. Data Analysis and Preliminary Results

Depending on their charge, different particles lose energy differently in the proportional counters. Thus, comparing the energy lost by the particles in the proportional counters ($\Delta E_1, \Delta E$) with the total energy deposited in the plastic scintillator (E), it was posible to identify the charged particles corresponding to different reaction channels. Fig. 4 shows the two-dimensional particle identification spectra that were used to identify the various reaction products. Accordingly, gates were set on the alphas to select focal plane spectra from the (d, α) channel alone. These spectra were used to obtain excitation energies and differential scattering cross sections. As shown in Fig. 3 we identified approximately 25 new states in ^{136}Cs below 2.5 MeV. The measured angular distributions were compared to the theoretical predictions from Distorted Wave Born Approximation (DWBA) calculations. These calculations, performed using the DWUCK4 code [15], were used in conjuction with the data to identify the orbital angular momentum (L) transferred for each observed excited state. The elastic scattering and the reaction cross section angular distributions were calculated using the optical model [16] to generate the distorted waves in the DWBA calculations. For these data we first used the global deuteron and alpha optical model parameters (OMPs) from [17,18] and [19] respectively. The OMPs were input into the DWUCK4 code to calculate the angular distributions. The measured elastic scattering cross sections at various angles were compared with the DWBA predictions to validate the parameters that would eventually be used for the ^{138}Ba(d, α) two-nucleon transfer cross sections. Based on this comparison, it was decided that the deuteron OMPs from [17] were better suited for our analysis. These parameters were used to obtain multiple angular distributions corresponding to various L-transfers for specific states of interest in ^{136}Cs with given angular momentum and parity. Fig. 5 shows the predicted angular distributions obtained for the ground state assuming various L values. On comparison with the measured distribution, one can easily assign a spin and parity of $J^{\pi} = 5^{+}$ for the ground state of ^{136}Cs, thus lending support to our calculations. A similar analysis for the excited states is currently in progress.

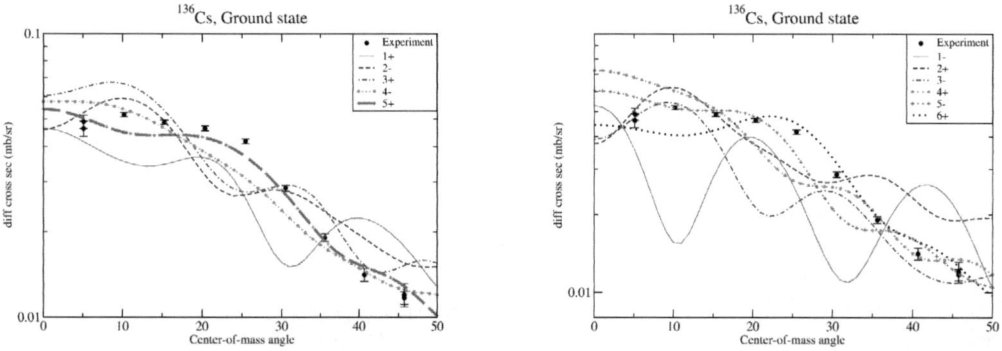

Figure 5. Measured angular distribution for the ground state (black points) of ^{136}Cs compared with the angular distributions obtained from DWUCK4 for different combinations of (L) transfers leading to states of natural parity $J^{\pi} = 1^{-}, 2^{+}, 3^{-}$, etc (right panel) and unnatural parity $J^{\pi} = 1^{+}, 2^{-}, 3^{+}$, etc (left panel). The different colors for the angular distribution curves obtained from DWUCK4 correspond to the spin-parity state resulting from the different L transfers. Natural parity states in a (d, α) reaction are populated by the $L = J$ transfer. For states of unnatural parity, two values $L = J \pm 1$ contribute for the same total angular momentum J. The spin-parity assignment of 5^{+} for the ground state of ^{136}Cs (left panel, bold, green dashed line) is made based on this comparison.

References

[1] Griffiths DJ *Introduction to Elementary Particles*, John Wiley & Sons.
[2] Gonzalez-Garcia MC and Maltoni M 2008 *Phys. Rept.* **460**, 1-129.
[3] Ahmad QR *et al.* 2002 *Phys. Rev. Lett* **89**, 011302.
[4] Vogel P 2012 *J. Phys. G: Nucl. Part. Phys.* **39**, 124002.
[5] Albert JB *et al.* 2014 *Nature* **510**, 229.
[6] Horoi M and Stoica S 2010 *Phys. Rev.* C **81**, 024321.
[7] Sen'kov R, Horoi M and Brown BA 2014 *Phys. Rev.* C **89**, 054304.
[8] Senkov RA and Horoi M 2013 *Phys. Rev.* C **88**, 064312.
[9] Ackerman N 2011 *Phys. Rev. Lett.* **107** 212501.
[10] KamLAND-Zen Collab. 2013 *Phys. Rev. Lett.* **110**, 062502.
[11] XMASS Collab. 2013 *J. Phys.: Conf. Series* **120** 042022.
[12] Puppe P *et al.* 2011 *Phy. Rev.* C **84**, 051305(R).
[13] Löffler L, *et al.* 1973 *Nucl. Instrum. Methods* **111**, 1.
[14] Wirth H-F *et al.* 2000. New Q3D Focal Plane Detector with Cathode-strip Readout Became Operational, Maier-Leibnitz-Laboratorium Jahresbericht, p.71.
[15] Kunz PD 1978 DWUCK4 DWBA Program, University of Colorado, Unpublished.
[16] Hodgson PE *Nuclear Reactions and Nuclear Structure*, Clarendon Press, Oxford
[17] Han Y, Shi Y and Shen Q 2006 *Phy. Rev.* C **74**, 044615.
[18] An H and Cai C 2006 *Phy. Rev.* C **73**, 054605.
[19] Burnett SM, Baxter AM, Hinds S and Pribac F 1985 *Nucl. Phys.* A**442**, 289-299.

Journal of Physics: Conference Series **689** (2016) 012027 doi:10.1088/1742-6596/689/1/012027

Effective band structure of Ru-doped $BaFe_2As_2$

M Reticcioli[1,2], G Profeta[2], C Franchini[1], A Continenza[2]

[1] Computational Materials Physics, University of Wien, Universitätsring 1, 1010 Wien (Austria)

[2] Department of Physical and Chemical Sciences, University of L'Aquila, Via Vetoio, 67100 Coppito (AQ) (Italy)

E-mail: michele.reticcioli@univie.ac.at

Abstract. The use of lattice cells in real space that are arbitrarily larger than the primitive one, is nowadays more and more often required by *ab initio* calculations to study disorder, vacancy or doping effects in real materials. This leads, however, to complex band structures which are hard to interpret. Therefore an unfolding procedure is sought for in order to obtain useful data, directly comparable with experimental results, such as angle-resolved photoemission spectroscopy measurements.

Here, we present an extension of the unfolding procedure recently implemented in the VASP code, which includes a projection scheme that leads to a full reconstruction of the primitive space. As a test case, we apply this newly implemented scheme to the Ru-doped $BaFe_2As_2$ superconducting compound. The results provide a clear description of the effective electronic band structure in the conventional Brillouin zone, highlighting the crucial role played by doping in this compound.

1 Introduction

The *ab initio* computational studies of systems with a broken lattice periodicity can be made easier using unfolding techniques. In fact, the unfolding is a procedure that enables to manage, and to easily interpret, the states folded into the momentum space of any supercell.

As a matter of fact, simulations often use supercells, rather than primitive cells, in order to handle the loss of translation symmetry of the lattice occurring in real systems. The break of periodicity can in fact arise every time one is dealing with some sort of disorder, such as systems with impurities (vacancies, dislocations, substitutions) or different spin configurations, as well as in case of doped materials and alloys. In all these cases, the use of a supercell allows to model random and disordered configurations by considering a periodic system having a larger unit cell volume.

However, using a supercell in real space implies a contraction of the corresponding unit cell in reciprocal space (i.e. the Brillouin zone), and hence the appearance of folded states. Theories of unfolding methods appeared in recent times and have already been computationally implemented and used for realistic applications [1–4]. In this work, we complement the unfolding procedure implemented in the VASP code [3, 5, 6] with a projection scheme for any vector in the primitive and supercell reciprocal spaces and apply our implementation to the iron based superconductor $BaFe_{2(1-x)}Ru_{2x}As_2$.

As well known [7] the pure iron-based pnictide families are not superconductors; however, substitution of other transition metals on Fe sites is able to drive these compounds into a

Journal of Physics: Conference Series **689** (2016) 012027 doi:10.1088/1742-6596/689/1/012027

superconducting state. In order to single out the mechanisms leading to superconductivity, it is therefore very important to study and understand the effects of the dopant on the electronic states of the pure compound. In these cases unfolding procedures are expected to be a convenient and powerful tool. In the following we describe the methodology (section 2) and then discuss its application to the $BaFe_{2(1-x)}Ru_{2x}As_2$ test case (section 3). Finally we will draw our conclusions.

2 The unfolding method

Our goal is to describe the energy band $E(\mathbf{K})$ obtained in the $\mathbf{K}$-vector reciprocal space related to the supercell in term of the $\mathbf{k}$-vector reciprocal space linked to the primitive cell. The result is a so called effective band structure (EBS) $E(\mathbf{k})$.

The connection between the primitive and supercell spaces is contained in the transformation matrix $\underline{M}$ (invertible and with integer elements $m_{ij} \in \mathbb{Z}$) defined by the following equation

$$\underline{\mathbf{A}} = \underline{M}\ \underline{\mathbf{a}} \quad \text{or, more explicitly,} \quad \begin{pmatrix} \mathbf{A}_1 \\ \mathbf{A}_2 \\ \mathbf{A}_3 \end{pmatrix} = \begin{pmatrix} m_{11} & m_{12} & m_{13} \\ m_{21} & m_{22} & m_{23} \\ m_{31} & m_{32} & m_{33} \end{pmatrix} \begin{pmatrix} \mathbf{a}_1 \\ \mathbf{a}_2 \\ \mathbf{a}_3 \end{pmatrix} \tag{1}$$

where $\underline{\mathbf{A}}$ and $\underline{\mathbf{a}}$ are the two sets of basis lattice vectors $\mathbf{A}_i$ and $\mathbf{a}_i$ (with $i = 1, 2, 3$) related to the super and primitive cell, respectively.

The unit cell vectors of the reciprocal space $\mathbf{B}_i$ and $\mathbf{b}_i$ are related by an analogous equation:

$$\underline{\mathbf{B}} = \left(\underline{M}^{-1}\right)^T \underline{\mathbf{b}}\ . \tag{2}$$

It should be underlined that in the original formulation of the EBS method [2] the matrix $\underline{\mathbf{B}}$ is given instead by $\underline{\mathbf{B}} = \underline{M}^{-1}\ \underline{\mathbf{b}}$. However, it can be easily proven, using the well known definition of reciprocal vector in terms of vectorial products of lattice vectors, that the latter equation is not valid for supercells associated with a non-symmetric transformation matrix $\underline{M}$. In this case and in general, Eq. 2, that includes the transpose operation, holds instead.

A wave vector belonging to the $\mathbf{k}$-space is said to be folded into a wave vector of the $\mathbf{K}$-space, if there exists a reciprocal lattice vector $\mathbf{G}_0$ such that $\mathbf{K} = \mathbf{k} - \mathbf{G}_0$. Likewise, a $\mathbf{K}$-vector unfolds to N $\mathbf{k}_i$-vectors defined by the following equation: $\mathbf{k}_i = \mathbf{K} + \mathbf{G}_i$ with $i = 1, \ldots, N$, where N is given by the determinant of the transformation matrix $\underline{M}$, representing the ratio of the two cells volumes, and $\mathbf{G}_i$ are the reciprocal lattice vectors obtained by linear combinations of the unit cell vector of the reciprocal space. Clearly, for a given $\mathbf{k}$-vector, there exists only a unique ($\mathbf{K}$, $\mathbf{G}_0$) pair fulfilling the previous relation, meaning that each $\mathbf{k}$-vector is folded into a single $\mathbf{K}$-vector. On the other hand, a $\mathbf{K}$-vector unfolds into N wave vector of the $\mathbf{k}$-space.

Any eigenvector $|\mathbf{K}m\rangle$ of the supercell can then be written in term of eigenvectors $|\mathbf{k}n\rangle$ of the primitive cell, since they form a complete basis set, where m and n stand for band indices. The probability $P_{\mathbf{K}m}(\mathbf{k}_i)$ relating the contribution of the $|\mathbf{k}_i n\rangle$ states to a $|\mathbf{K}m\rangle$ state is given by the following equation

$$P_{\mathbf{K}m}(\mathbf{k}_i) = \sum_n |\langle \mathbf{K}m|\mathbf{k}_i n\rangle|^2\ , \tag{3}$$

that defines also a spectral function $A(\mathbf{k}_i, E)$:

$$A(\mathbf{k}_i, E) = \sum_m P_{\mathbf{K}m}(\mathbf{k}_i)\ \delta(E_m - E) = \sum_m \sum_n |\langle \mathbf{K}m|\mathbf{k}_i n\rangle|^2\ \delta(E_m - E)\ . \tag{4}$$

The quantity $P_{\mathbf{K}m}(\mathbf{k}_i)$ represents the amount of Bloch character of $|\mathbf{k}_i n\rangle$ preserved in $|\mathbf{K}m\rangle$ and is fundamental to calculate the effective band structure, that is the band structure $E(\mathbf{k})$ of the primitive cell calculated starting from the dispersion relation $E(\mathbf{K})$ of the supercell. In

fact, the Bloch character of each $|\mathbf{K}m\rangle$ state allows to unfold these states into the $\mathbf{k}$-space of the primitive cell, since the eigenvalue $E(\mathbf{K})$ related to a $|\mathbf{K}m\rangle$ eigenstate is also related to the $|\mathbf{k}_i n\rangle$ states with a probability $P_{\mathbf{K}m}(\mathbf{k}_i)$ different from zero.
If the supercell is a mere spatial repetition of identical primitive cells (i.e. same lattice vectors and same atomic basis), then there is only one $|\mathbf{k}_i n\rangle$ state contributing to $P_{\mathbf{K}m}(\mathbf{k}_i)$ for each $|\mathbf{K}m\rangle$ state. All other contributions to $P_{\mathbf{K}m}(\mathbf{k}_i)$ will vanish. In this case, the effective band structure calculated starting from the eigenvalues of a supercell is an exact reconstruction of the band structure of the primitive cell. Due to this perfect equivalence, there is no need to use the unfolding procedure.

On the other hand, the unfolding method becomes useful when the supercell is constructed with unit cells having the same lattice parameters (in order to maintain the relation $m_{ij} \in \mathbb{Z}$ valid), but with different atomic basis. In this situation, many different $|\mathbf{k}_i n\rangle$ will contribute to $P_{\mathbf{K}m}(\mathbf{k}_i)$ for a single $|\mathbf{K}m\rangle$. This is due to the fact that the $\mathbf{k}$-vectors are not good quantum quantities anymore, since translational symmetry has been broken. In fact, translational symmetry in this case survives only in the $\mathbf{K}$-space related to the supercell. The application of the unfolding scheme in this case leads to an effective band structure with broadened bands.

In the plane waves basis set used by the VASP code the Bloch character $P_{\mathbf{K}m}(\mathbf{k})$ can be conveniently calculated starting from eigenvectors belonging to the supercell only, thus avoiding the calculation of the eigenvectors of the primitive cell, as shown in the following equation [2]:

$$P_{\mathbf{K}m}(\mathbf{k}) = \sum_{\{\mathbf{g}\}} |C_{m,\mathbf{g}+\mathbf{k}}|^2 , \tag{5}$$

where $\{\mathbf{g}\}$ is the set of reciprocal vectors of the primitive cell and $C_{m,\mathbf{g}+\mathbf{k}}$ is the coefficient for the $|\mathbf{K}m\rangle$ eigenstate in a plane wave basis set expansion. The Bloch character coefficients $P_{\mathbf{K}m}(\mathbf{k})$ are provided for the full set of N primitive $\mathbf{k}$-points $\mathbf{k}_i = \mathbf{K} + \mathbf{G}_i$. However, since the analysis of the band structure is usually restricted to a limited set of primitive $\mathbf{k}$-points, it would be useful to restrict the EBS scheme to such a set. Starting from Eqs. 1 and 2, we have implemented an efficient projection scheme of the $\mathbf{K}$-points of the supercell into a limited set of $\mathbf{k}$-points in the primitive one, freely chosen by the user. The folded $\mathbf{K}$-points in the supercell are calculated automatically and then used internally in the code to calculate eigenvalues and Bloch character coefficients to obtain band structure plots after self-consistency is achieved, as usually done in ordinary band structure calculations.

Therefore, it is possible to obtain an effective band structure starting from a given supercell, the lattice vectors of a primitive cell and a chosen set of $\mathbf{k}$ points, simplifying the complexity of the unfolding procedure.

3 Test case: $\mathbf{BaFe_{2(1-x)}Ru_{2x}As_2}$

We have tested our implementation on the $BaFe_{2(1-x)}Ru_{2x}As_2$ compound. As mentioned before, transition metal doping in these compounds is able to induce superconductivity at certain optimal, usually small, concentrations [8, 9]. As the dopant concentration is increased, the impurity scattering brought about the dopant atoms, usually destroys the superconducting state. Ru doped $BaFe_2As_2$ compounds are a rather peculiar case since they exhibit a persistent resilience of superconductivity up to quite large concentrations of disordered impurities (about 40% Ru-substitution [10–13]). Then, at larger concentrations, the critical temperature for the onset of the superconducting state is seen to progressively decrease.

3.1 Computational setup

The calculations have been performed using the VASP [5,6] package and projected augmented-wave (PAW) [14] pseudopotentials, within the generalized gradient approximation [15] (GGA) to the density functional theory, since this approximation is seen to better reproduce the main structural features of the Fe-based pnictides [16]. For the iron atoms, inclusion of p-electrons in the valence shell has been found to be relevant for a more accurate description of the system.

The isovalent substitution of Ru-atoms on Fe-sites was investigated using a supercell containing 16 Fe/Ru atoms, 8× larger than the 2-Fe primitive cell of Fig. 1(a), with a

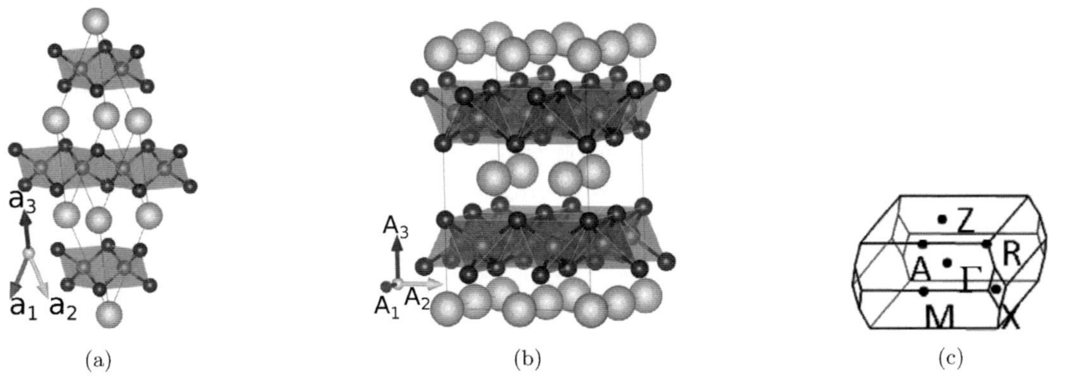

Figure 1. $BaFe_2As_2$ structure: fig. 1(a) and 1(b) represent the primitive cell and the supercell respectively used in the current work (Ba-atoms are represented in green color, Fe in brown, As in blue). Bands are unfolded on the Brillouin zone relative to the primitive cell shown in fig. 1(c).

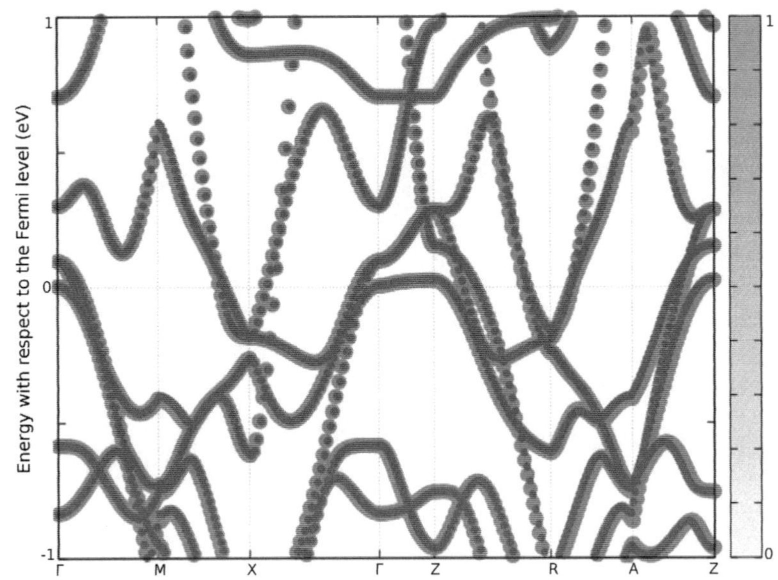

Figure 2. Effective band structure (gradient of yellow) compared with the primitive cell (blue). The intensity of the unfolded bands represents the Bloch character $P_{\mathbf{K}}(\mathbf{k}_i)$ normalized with respect to the contributions belonging to all the $\mathbf{k}_i$-points connected by a $\mathbf{G}_i$-vector to the $\mathbf{K}$-point of the supercell. In this case the supercell used is without any impurity (i.e. $x = 0$). Since it contains no impurities, the Bloch character is either one or zero for every $\mathbf{k}$-point.

transformation matrix $\underline{M}$ (see eq. 1) defined by the following relation:

$$\underline{M} = \begin{pmatrix} 0 & 2 & 2 \\ 2 & 0 & 2 \\ 1 & 1 & 0 \end{pmatrix} \text{ implying } \begin{cases} \mathbf{A}_1 = 2\,\mathbf{a}_2 + 2\,\mathbf{a}_3 \\ \mathbf{A}_2 = 2\,\mathbf{a}_1 + 2\,\mathbf{a}_3 \\ \mathbf{A}_3 = \mathbf{a}_1 + \mathbf{a}_2 \end{cases} . \tag{6}$$

We considered the primitive cell with experimental lattice parameters and let the internal atomic positions relax, following the *ab initio* calculated forces, performing spin-polarized calculations. Self consinstency and convergence of the relaxed internal structural parameters were reached within a variation for the total free energy of 10^{-4} eV and 10^{-3} eV respectively, using 500 eV as energy cutoff. Integration of the irreducible Brillouin zone was performed considering $8\times\ 8\times\ 3$ grids within the Monkhorst and Pack scheme [17]. Band structure eigenvalues were calculated on a finer grid (40 points per line) in the Brillouin zone. Then, we obtained the EBS by unfolding the band structure into the 2-Fe primitive cell of Fig. 1 using the improved unfolding scheme described above.

3.2 Results

In order to perform the test, a small (2×) supercell without any impurities ($x = 0$) was built: the EBS of this pure supercell, being simply a spatial repetition of identical primitive cells, should give exactly the same eigenvalues as those obtained from an ordinary band structure calculation of the pure $BaFe_2As_2$ primitive cell. Figure 2 shows the good agreement between the effective and ordinary band structure. Due to the perfect translational symmetry of the primitive cells

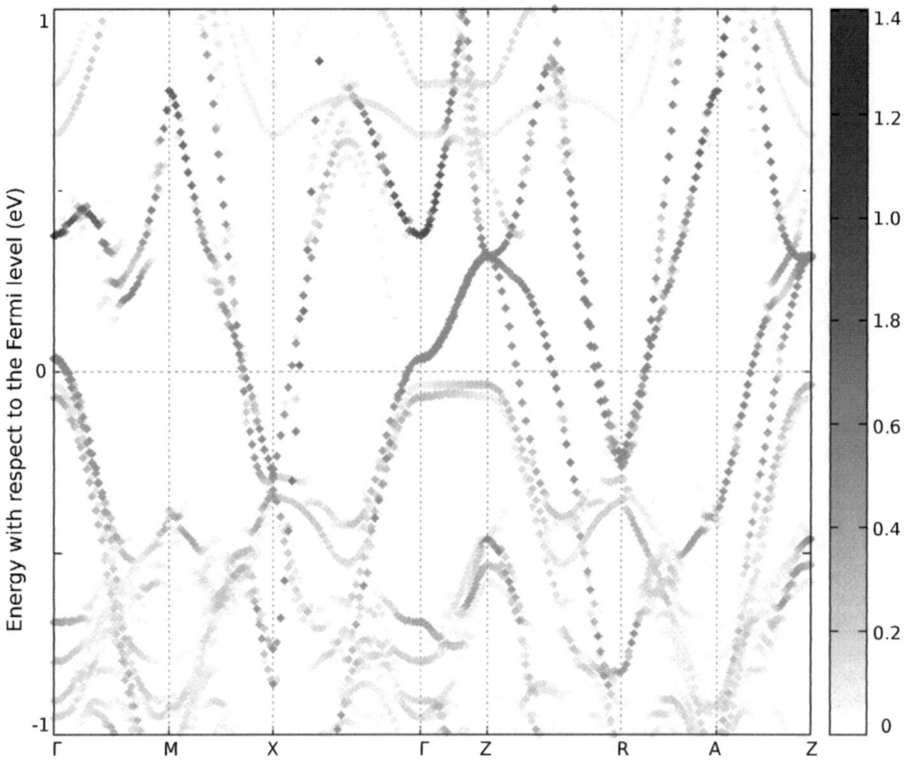

Figure 3. Effective band structure of $BaFe_{2(1-x)}Ru_{2x}As_2$ compound with $x = 0.5$. The intensity of the unfolded eigenvalues represents the Bloch character $P_{\mathbf{K}}(\mathbf{k}_i)$.

building the supercell system the bands are sharply resolved, i.e. the Bloch character is equal to one or zero for each point in the band structure, with vanishing probability for intermediate values. In this case in fact, only a single eigenstate of the primitive cell will contribute to the eigenstate of the supercell reciprocal **K**-space.

The effect of Ru-dopant atoms on the $BaFe_{2(1-x)}Ru_{2x}As_2$ compound have been studied for several concentration levels. Here we provide a description of the difference between the undoped and the 50% doped compound, in order to highlight the effect of the Ru-substitution. The results shown in Fig. 3 for $x = 0.5$ Ru-concentration make evident that the Ru-Fe substitution, even if it is usually considered to be isovalent, deeply modifies the electronic structure at the Fermi level. In particular, the hole pockets around the Γ-point in the Brillouin zone, shrinks with respect to the pure case (shown in Fig. 2) as the Ru-concentration is increased. On the other hand, the electron pocket at the X corner of the zone remains unchanged. An other effect of the Ru substitution is a broadening of the band structure. However, the bands appear to be broadened only in a region far from the Fermi level. Only the already described lowering of some bands affects the states at the Fermi level.

4 Conclusions

We discussed an improved unfolding procedure to calculate the effective band structure of supercells. The unfolding method implemented within the VASP code and here improved, has been applied to the $BaFe_{2(1-x)}Ru_{2x}As_2$ compound. The unfolding technique allows us to study the changes of the band structure upon Ru-substitution by means of a direct comparison with the pure compound. We have shown that the isovalent Ru-substitution causes the shrinking of the hole pockets around the Γ-point of the primitive cell, in good agreement with previous calculations [4]. These changes lead to modifications of the Fermi surface topology that might affect the pair coupling mechanisms. This procedure is general and can be applied to any compounds. We plan to apply this methodology for a systematic study of the role of dopant atoms in the superconducting properties of iron based pnictides.

Acknowledgments

This work has been carried out within a joint collaboration between the University of L'Aquila (Italy) and the University of Vienna (Austria), supported by the LLP Erasmus Placement student mobility program and the FWF-SFB ViCoM (Grant No. F41) project. Computing time at the Vienna Scientific Cluster and ISCRA C grant at Cineca HPC are greatly acknowledged.

References

[1] Popescu V and Zunger A 2010 *Phys. Rev. Lett.* **104** 236403
[2] Popescu V and Zunger A 2012 *Phys. Rev.* B **85** 085201
[3] Eckhardt C Hummer K and Kresse G 2014 *Phys. Rev.* B **89** 165201
[4] Wang L *et al.* 2013 *Phys. Rev. Lett.* **110** 037001
[5] Kresse G and Furthmüller J 1996 *Phys. Rev.* B **54** 11169
[6] Kresse G and Furthmüller J 1996 *Comp. Mat. Sc.* **6** 15
[7] Kamihara Y *et al* 2008 *J. of the Am. Chem. Soc.* **130** 3296
[8] Thaler A *et al.* 2010 *Phys. Rev.* B **82** 014534
[9] Kim M G *et al.* 2011 *Phys. Rev.* B **83** 054514
[10] Dhaka R S *et al.* 2011 *Phys. Rev. Lett.* **107** 267002
[11] Brouet V *et al.* 2010 *Phys. Rev. Lett.* **105** 087001
[12] Rullier-Albenque F *et al.* 2010 *Phys. Rev.* B **81** 224503
[13] Xu N *et al.* 2012 *Phys. Rev.* B **86** 064505
[14] Blochl P L 1994 *Phys. Rev.* B **50** 17953
[15] Perdew J P *et al.* 1996 *Phys. Rev. Lett.* **77** 3865
[16] Mazin I I *et al.* 2008 *Phys. Rev.* B **78** 085104
[17] Monkhorst H J and Pack J D 1976 *Phys. Rev.* B **13** 5188

Journal of Physics: Conference Series **689** (2016) 012028 doi:10.1088/1742-6596/689/1/012028

The role of metals in protein conformational disorders - The case of prion protein and Aβ-peptide

E De Santis[1], V Minicozzi[1], S Morante[1], G C Rossi[1,2] and F Stellato[1]

[1]Dipartimento di Fisica, Università di Roma Tor Vergata and INFN Sezione di Roma Tor Vergata, Via della Ricerca Scientifica, 00133, Roma, Italia
[2]Centro Fermi, Piazza del Viminale, 00184, Roma, Italia

E-mail: edesantis@roma2.infn.it

Abstract. Protein conformational disorders are members of a vast class of pathologies in which endogenous proteins or peptides undergo a misfolding process by switching from the physiological soluble configuration to a pathological fibrillar insoluble state. An important, but not yet fully elucidated, role in the process appears to be played by transition metal ions, mainly copper and zinc. X-ray absorption spectroscopy is one of the most suitable techniques for the structural characterization of biological molecules in complex with metal. Owing to its chemical selectivity and sensitivity to the local atomic geometry around the absorber, it can be successfully used to study the environment of metal ions in complex with proteins and peptides in physiological conditions. In this paper we present X-ray absorption spectroscopy studies of the metal ions coordination modes in systems where metals are complexed with specific amyloidogenic proteins and peptides. In particular, we show results concerning the Amyloid β peptide, that is involved in Alzheimer's disease, and the Prion protein, that is responsible for the Transmissible Spongiform Encephalopathy. Our findings suggest that the copper and zinc ions may play a crucial role in the aggregation and fibril formation process of these two biomolecules. Elucidating this kind of interaction could be a key preliminary step before any viable therapy can be conceived or designed.

1. Introduction

It is a well established notion that the function of a protein is strictly related to its structure. Protein folding is a very complicated process that not always ends with the 'right' three-dimensional configuration. Being protein function strictly related to its folding, it is straightforwardly understood how errors in the folding can be at the basis of pathologies. Indeed, protein conformational disorders are pathologies caused by the transition of endogenous proteins and peptides from the physiological globular configuration to a misfolded harmful fibrillary state. Regardless of the nature of the proteins by which they are formed, fibrils have common histochemical properties and ultrastructural morphology.

Among this class of pathologies, many neurodegenerative diseases have a major impact on human health, in particular, Transmittable Spongiform Encephalopathies (TSEs), Alzheimer's (AD) and Parkinson's (PD) diseases. The molecular basis of these diseases is not yet completely understood. An important, not yet fully elucidated, role appears to be played by transitions metal ions (mainly Cu^{2+} and Zn^{2+}) that have been observed to be present in fairly large amount in the aggregates. The presence of these metal ions in amyloid fibrils evokes the possibility that they might trigger, prevent or promote amyloid aggregates formation [1, 2].

In this general framework, it appears to be of the utmost importance to understand and clarify whether and how metal ions cross-interact with aggregating proteins and peptides. Here, we present the analysis, at a structural level, of the mechanism by which Cu^{2+} and Zn^{2+} ions appear to compete in the binding to both the prion protein, which is involved in TSEs, and to the Aβ peptides, which is involved in AD.

1.1. Transmittable Spongiform Encephalopathies (TSEs) and Prion protein (PrP)

TSEs are fatal infectious neurodegenerative diseases that affect several animal species. Examples of TSEs include mad cow disease in cattle, scrapie in goats and sheep, and Kuru and Creutzfeldt–Jakob diseases in humans. All TSEs are characterized by spongiform degeneration (formation of fluid-filled holes in the brain tissue) and astrocytic gliosis (proliferation and branching of glial) of the central nervous system that unavoidably lead to death. There is increasing evidence that in all TSEs the occurrence of an abnormal form of a protein called Prion protein (contraction of "proteinaceous infectious protein" [3]) is of crucial importance. The PrP is a membrane protein composed by about 260 amino acids anchored to the membrane surface via a glycosyl-phosphatidyl inositol group. It exists in two alternative conformers: the cellular native conformer, PrP^C, rich in α-helix, and the aberrant pathogenic conformer, PrP^{SC} (or scrapie PrP) endowed with aberrant self-replicating properties, comparatively rich in β-sheet [4]. Although the physiological role of PrP^C is not known, the ability of PrP to bind both Cu^{2+} and Zn^{2+} ions *in vivo* has led to the suspicion that this protein may be involved in metal homeostasis in the brain [5-8]. The observation that Cu and Zn homeostasis is severely altered in the brain tissues of infected organisms compared with normal tissues led to the hypothesis of an involvement of these two metals in the illness genesis.

From a structural point of view, the N-terminal region of PrP, up to residue 120, is unstructured and highly flexible. A characteristic feature of this region is the so-called octarepeat domain composed by a number of repeats (up to six, depending on the species) of the eight-residue sequence PHGGGWGQ. It has been shown that this region hosts the binding sites with the highest affinity to metals [6, 9] (see Figure 1a).

Electron Paramagnetic Resonance (EPR) experiments, performed on hamster PrP, have shown that the copper coordination environment depends critically on the [Cu^{2+}]:[octarepeat] concentration ratio. At low Cu^{2+} concentration, the four octarepeat His imidazole side chains bind simultaneously to a single Cu^{2+} [10]. At intermediate Cu^{2+} concentration, the octarepeat takes up two copper equivalents, each coordinated by two His side chains. At higher copper concentration, the octarepeat domain saturates at 4 copper equivalents [11]. It has been also demonstrated [5, 11] that Zn^{2+} is able to perturb the Cu^{2+} coordination mode to the tetra-octarepeat region. In particular, it has been suggested that, even if Zn^{2+} is not able to completely remove Cu^{2+}, by increasing its concentration the way Cu^{2+} is bound to the tetra-octarepeat can be progressively changed. The effect of Zn^{2+} on Cu^{2+} binding is stronger at moderate Cu^{2+} concentration and progressively disappears while Cu^{2+} concentration is increased.

1.2. Alzheimer disease (AD) and Amyloid β peptide (Aβ)

AD is a type of dementia that causes problems with memory, thinking and behavior. Symptoms usually develop slowly and get worse over time, becoming severe enough to interfere with daily tasks. The brain of people affected by AD shows the presence of two microscopic kind of aggregates: the senile (or amyloid) plaques, which are found between neurons, and the neurofibrillary tangles, which are found inside neurons. Amyloid plaques are formed by the aggregation of Aβ. Aβ is originated by the proteolytic cleavage of a membrane protein called Amyloid Precursor Protein (APP) that is encoded in chromosome 21. Under physiological conditions, various secretases are known to cut the APP. If α and γ secretases are involved in the cutting, the result is a non-pathological, harmless Aβ-peptide form. However, if β and γ secretases are involved, the result is a pathological, harmful Aβ-peptide form. The formation of amyloid plaques from the peptides is a complex and not yet fully elucidated process. At the basis of this aggregation process there is a conformational change of the peptide that leads to the formation of an extended β-sheet like structure, capable of grouping together other peptides to form antiparallel β-sheet with intermolecular binding. It has been observed that plaques contain large amounts of transition metal ions (Cu^{2+} and Zn^{2+} are the most abundant). The physiological role possibly played by metal ions has not been yet fully understood. It has been shown [12], however, that the nature of the metal binding can influence the protein folding/misfolding and aggregation process, thus possibly affecting the pathology progression.

In Aβ the highest affinity metal binding sites are located in the first sixteen amino acids of the N-terminal fragment [13, 14] (see Figure 1b). Most of the studies in literature state that copper has an intra-peptide coordination involving the imidazole groups of three histidines (His_6, His_{13}, His_{14}). The nature of other amino acids possibly involved in Cu^{2+} coordination is still undefined. In many EPR works [15-17] it has been shown that Cu^{2+} coordination is heterogeneous at physiological pH. Namely two binding ways coexist and are termed Component I and Component II depending on the number of histidine residues bound to the metal (two His in Component I and one His in Component II).

The Zn^{2+} binding mode is more controversial and a few different structures in which Zn^{2+} is coordinated to different numbers of His residues have been proposed [18, 19]. Among them, one of the most accredited is an inter-peptide coordination mode in which Zn^{2+} is bound to four histidine residues [13, 20]. The simultaneous presence of both metal ions, that is the normal situation *in vivo*, is still poorly investigated. A multi techniques study of Damante *et al.* [21] suggested that the addition of Zn^{2+} does not promote Cu^{2+} ions release, but it significantly modifies the Cu^{2+} mode of coordination in Aβ. EPR studies of Silva *et al.* [15] have recently shown that after the addition of Zn^{2+} to the already formed Cu-complexes, Zn^{2+} is able to replace only Cu^{2+} coordinated in Component I, while Component II bound Cu^{2+} remain unmodified.

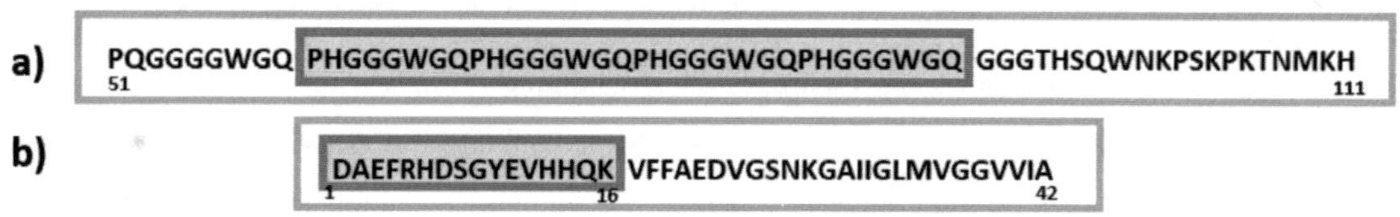

Figure 1 Primary structure of the region of PrP involved in metal binding (*panel a*) and of Aβ peptide (*panel b*). Regions with the highest binding probability are highlighted.

2. XAS – X-ray absorption spectroscopy

Owing to its chemical selectivity and sensitivity to local atomic arrangement around the absorber, X-ray absorption spectroscopy (XAS) can be successfully used for structural studies on biological materials, in particular when one is interested in characterizing the structural environment of metal ions (that are the absorbers in the systems studied in this paper) complexed with proteins and peptides in physiological conditions. X-ray photons used for XAS experiment are usually produced by synchrotrons, which provide intense and tunable X-ray beams. The X-ray beam is then focused on the samples and the structural information about the atomic environment around the absorber are obtained through the study of the absorption coefficient as a function of the photon energy. The absorption spectrum (XAS), whose oscillation originate from the interference between the outgoing and the backscattered electron waves, contains detailed information about the geometrical distribution of the scatterers around the absorber. The XAS spectrum is normally divided in two regions, the XANES (X-ray Absorption Near Edge Structure) and the EXAFS (Extended X-ray Absorption Fine Structure) (see Figure 2). XANES, where the kinetic energy of the ionized electron is low, is dominated by multiple scattering events, while the EXAFS region, that conventionally starts about 100 eV after the edge (i.e. the ionization energy of the 1s electron of the absorbing atom) is dominated by single scattering events. Compared to the EXAFS region, XANES are richer in geometrical information that, however, are much more difficult to extract.

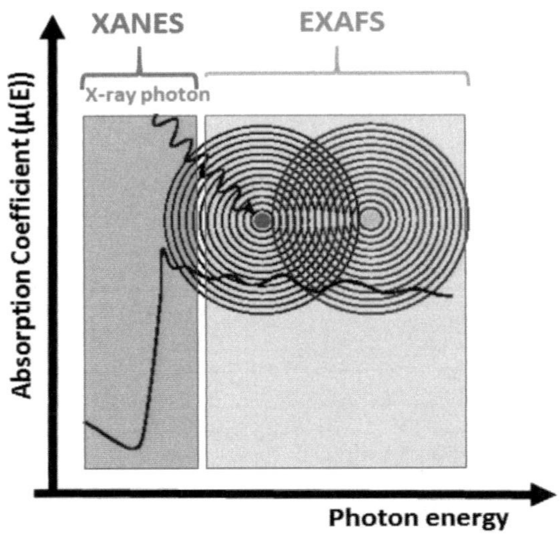

Figure 2 Schematic view of XAS oscillations due to the photoelectron scattering from neighbour atoms. XANES and EXAFS regions are highlighted.

3. Results and Discussion

Metal homeostasis is necessary to maintain essential transition metal ions at the required cellular concentration to both prevent their excess and their lack. There are evidences that a breakdown in metal trafficking regulation has a significant impact in the development of age-related neurodegenerative diseases [22]. PrP proteins and Aβ peptides are both present in brain. They display different affinity for specific metal ion binding and may possibly interact one with the other [23, 24]. Taken together, these three observations lead to the hypothesis that the two peptides may play complementary roles in the Cu^{2+} and Zn^{2+} ions homeostasis.

In the past years, many experimental techniques have been employed to characterize the structure of metal binding site in amyloid peptides, among which EPR [11, 15, 16, 21, 25], Nuclear Magnetic Resonance (NMR) [18, 19, 26], Circular Dichroism (CD) [25-27] and XAS [4-5, 12, 13, 28-31] are the most prominent ones. Using XAS, we investigated the way Cu^{2+} and Zn^{2+} reciprocally affect the way they bind to both PrP and Aβ peptide. Structural information was obtained both from the XANES and the EXAFS regions of the spectrum.

3.1. Prion Protein (PrP)

Systematic XAS experiments were performed on samples containing the tetra-octarepeat ($4R_8$) region of the PrP in the presence of Cu^{2+} and Zn^{2+}. In order to explore the possibility that the way the two metal ions compete and interfere in protein binding may also depend on the order they are added to the peptide containing solution, we prepared solutions where the Cu^{2+} (Zn^{2+}) ions were added after few hours of Zn^{2+} (Cu^{2+}) incubation in PrP solutions [5, 31].

The first result obtained through the EXAFS analysis is that both Cu^{2+} and Zn^{2+} bind to the $4R_8$ region via histidine residues when separately added, thus they compete for the same binding site. The second result is that Zn^{2+} binding mode changes upon Cu^{2+} addition to accommodate the copper ion. On the contrary, when Zn^{2+} is dissolved in a solution pre-incubated with Cu^{2+}, it does not modify the way Cu^{2+} binds the protein. The XAS experimental outcome is pictorially schematized in Figure 3.

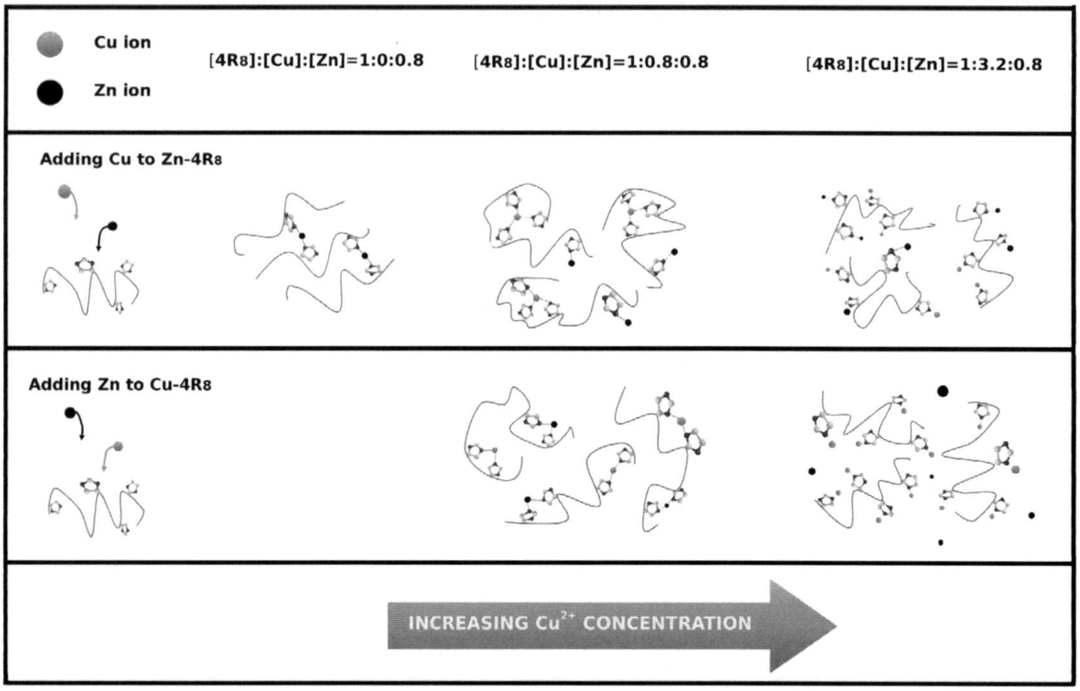

Figure 3 Schematic illustration of the suggested effect of adding Cu^{2+} and Zn^{2+} to the $4R_8$ solution in different order. Only the bonds between metals and His residues are drawn as short blue segments. In the first line, we depict the effect of adding Cu^{2+} to already formed $4R_8/Zn^{2+}$ complexes. In the second one, we show the effect of adding Zn^{2+} to already formed $4R_8/Cu^{2+}$ complexes. In the different panels, the number of orange (Cu^{2+}) and black (Zn^{2+}) dots per $4R_8$ is indicative only and does not represent the real proportion among tetra-octarepeats, Cu^{2+} and Zn^{2+}.

3.2. Aβ-peptide

XAS has been also used to study the metal binding mode to the Aβ peptide, showing that it depends on the nature and concentration of the complexed metal ion [32]. On one hand, XAS data are always consistent with an intra-peptide (involving a single copy of the Aβ peptide) Cu^{2+} mode of binding involving three His residues coordinated to the ion. On the other hand XAS data are consistent with either an intra- or an inter-peptide Zn^{2+} mode of binding, depending on solution conditions [12, 30].

As in the case of PrP, an obvious extension of this work is to study the way the two metals influence reciprocally their mode of binding to the Aβ peptide when they are simultaneously present. This study has already been performed using EPR that, however, cannot give information on the Zinc ion [15]. It is also worth studying the effect of the order in which metals are added to the peptide solution [33].

3.3. Conclusions

The general idea that emerges from the studies presented here is that both PrP and Aβ peptide may be involved in metal trafficking and homeostasis. The role played by metals in the pathological misfolding and aggregation processes is not yet clarified, but there are convergent evidences that Zn^{2+} play an important role at least in the first phases of aggregation. This role is possibly modulated by Cu^{2+} presence.

We are confident that, given the fact that the resolution of experimental techniques is constantly getting higher and higher and the parallel computing platforms is allowing to perform simulations of larger and larger (more detailed) model systems, we will collect enough information to at least clarifying the structural relevant steps of the misfolding-aggregation process involving metal ions.

Acknowledgments

We would to thank the 6th Young Researcher Meeting organizers and the GSSI institute for the hospitality and helpfulness. This work was supported by SUMA-INFN (Italy).

References

[1] Viles J H 2012 *Coord. Chem. Rev.* **256** 2271-2284
[2] Faller P *et al* 2013 *Inorg. Chem.* **52.21** 12193-12206
[3] Prusiner S B *et al* 1982 *Science* **216** 136-144
[4] Morante S 2008 *Curr. Alz. Res.* **5(6)** 508-524
[5] Stellato F *et al* 2014 *Eur. Biophys. J.* **43** 631-642
[6] Millhauser G L 2007 *Annu. Rev. Phys. Chem.* **58** 299
[7] Kenche V B *et al* 2011 *Br. J. Farmacol.* **163(2)** 211–219
[8] Pushie M J *et al* 2011 *Metallomics* **3.2** 206-214
[9] Morante S *et al* 2004 *J. Biol. Chem.* **279(12)** 11753-9
[10] Chattopadhyay M *et al* 2005 *JACS* **127.36** 12647-12656
[11] Millhauser G L 2013 *Springer New York* 21-34
[12] Morante S *et al* 2014 *Adv. Alzh. Res.* **2** 100-147
[13] Minicozzi V *et al* 2008 *J. Biol. Chem.* **283(16)** 10784-92
[14] Tsvetkov P O *et al* 2010 *Biophys. J.* **99.10** L84-L86
[15] Silva K I *et al* 2013 *J. Phys. Chem. B* **117** 9386-9394
[16] Silva K I *et al* 2014 *J. Phys. Chem. B* **118** 8935-8944
[17] Shin B K et al 2011 *J. Phys. Chem. A* **115** 9590–9602
[18] Danielsson J *et al* 2007 *FEBS J.* **274.1** 46-59
[19] Syme C D *et al* 2006 *Biochim. Biophys. Acta.* **1764** 246-256
[20] Giannozzi P *et al* 2012 *Metallomics* **4.2** 156-165
[21] Damante C A *et al* 2011 *Inorg. Chem.* **50** 5342–5350
[22] Barnham K J *et al* 2004 *Nat. Rev. Drug. Disc.* **3** 205-214
[23] Hernandez-Rapp J *et al* 2014 *Front. Aging. Neurosci.* **6**
[24] Laurén J *et al* 2009 *Nature* **457** 1128-1132
[25] Syme C D *et al* 2004 *J. Biol. Chem.* **279** 18169-18177
[26] Ma Q F *et al* 2006 *Biopolymers* **83** 20-31
[27] Minicozzi V *et al* 2014 *J. Bio. Chem.* **289(16)** 11242-52
[28] Dong J *et al* 2006 *J. Am. Chem. Soc.* **128** 3540-3542
[29] Alies B *et al* 2013 *Chem. Comm.* **49** 1214-1216
[30] Stellato F *et al* 2006 *Eur. Biophys. J.* **35** 340-351
[31] Stellato F *et al* 2011 *Eur. Biophys. J.* **40** 1259-1270
[32] Minicozzi V *et al* 2008 *Int. J. Quantum Chem.* **108.11** 1992-2015
[33] De Santis E *et al* 2015 *J. Phys. Chem. B* **accepted**

Monte Carlo studies for medical imaging detector optimization

G R Fois[1,2], E Cisbani[2], F Garibaldi[1,2]

[1] Istituto Nazionale di Fisica Nucleare (INFN), sezione di Roma, Piazzale Aldo Moro 2, 00181 Roma

[2] Istituto Superiore di Sanità (ISS), viale Regina Elena 299, 00161 Roma

E-mail: giovannarosa.fois@gmail.com

Abstract. This work reports on the Monte Carlo optimization studies of detection systems for Molecular Breast Imaging with radionuclides and Bremsstrahlung Imaging in nuclear medicine.

Molecular Breast Imaging requires competing performances of the detectors: high efficiency and high spatial resolutions; in this direction, it has been proposed an innovative device which combines images from two different, and somehow complementary, detectors at the opposite sides of the breast. The dual detector design allows for spot compression and improves significantly the performance of the overall system if all components are well tuned, layout and processing carefully optimized; in this direction the Monte Carlo simulation represents a valuable tools.

In recent years, Bremsstrahlung Imaging potentiality in internal radiotherapy (with beta-radiopharmaceuticals) has been clearly emerged; Bremsstrahlung Imaging is currently performed with existing detector generally used for single photon radioisotopes. We are evaluating the possibility to adapt an existing compact gamma camera and optimize by Monte Carlo its performance for Bremsstrahlung imaging with photons emitted by the beta- from ^{90}Y.

1. Introduction

Medical imaging is a field of research in continuous development. This sector is of fundamental importance for the early diagnosis and treatment of many diseases. Medical imaging includes a whole series of techniques, very different from each other, whose common purpose is to make visible the interior structures of the body. Nowadays various techniques are in use, such as: radiography, which is among the most common tests for the study of the skeletal system, the CT (Computed Tomography) that allows to obtain three-dimensional images. In the field of nuclear medicine is worth cite the PET (Positron Emission Tomography) which provides physiological information through maps of functional processes within the body. There are also a number of imaging techniques that do not make use of ionizing radiation such as ultrasound, which exploits the transmission of the ultrasonic waves, and MRI (Magnetic Resonance Imaging), which is a technique of image generation based on the physical principle of nuclear magnetic resonance.

In this broader context we focused our attention on two molecular imaging techniques based on single photon detection: MBI (Molecular Breast Imaging) and BSI (Bremsstrahlung Imaging).

1.1. MBI
Breast cancer is one of the most common cancers in women. Early detection and improved diagnosis lead to more efficient treatment and better outcomes. Mammography is the current gold standard for early breast cancer detection. Mammography is typically very sensitive but not specific. In addition, his sensitivity is significantly reduced in certain subset of women, particularly in women with radiographically dense breasts. As a result, majority of mammography-directed biopsies are performed on benign lesions and a significant number of biopsies are therefore unnecessary. Recognizing the limitation of mammography, additional adjunct diagnostic methods need to be applied. Among them molecular imaging with radionuclides has a central role to play.

MBI is a nuclear medicine technique that provide high-resolution functional images of the breast. MBI have as their origin an older technique called scintimammography which used conventional γ-cameras to image the uptake of Tc-99m sestamibi in breast tumors. Because of the large dead space at the edge of the camera requiring the patient to be imaged prone with the camera positioned laterally, the resulted distance between the breast and the camera increased leading to a considerable loss of resolution [1]. So the standard γ-cameras initially used do not offer sufficient spatial resolution to detect small lesions when used in breast imaging scintimammography procedure. O'Connor et al. [2] studied a dual-head system which allows simultaneous acquisition of superior and inferior views of the breast. Thanks to the use of such dual-head MBI system, a significantly increase sensitivity for subcentimeter lesions was found.

Our group has designed and implemented a dual detector layout which consists of two detectors of different dimensions: a 150×200 cm^2 with parallel hole collimator the first, and a 7×7cm^2 with pinhole collimator the second. This asymmetric setup allows monitoring the whole breast and performing spot compression bringing the detector closer to the potential suspicious lesion therefore increasing the efficiency and at the same time imaging it with relatively high resolution thanks to the pin-hole magnification [3].

1.2. BSI
The effective use of radionuclides in cancer treatment such as TRT (Targeted Radionuclide Therapy) requires the knowledge of their spatial and temporal distribution on a personalised base. The use of beta-emitters radionuclides has been proposed as a means of enhancing the dose deposition within the targeted cells while minimizing the dose received by other organs. The Bremsstrahlung interactions of the emitted beta in the body produces a continuous spectrum of photons that can be used for imaging the beta-radiopharmaceuticals distribution. [4].

Our work aims to investigate the Bremsstrahlung imaging with compact gamma cameras and identify the best procedures and solutions to maximise its quality.

2. Materials and methods
This Monte Carlo study has been carried out using the GATE simulation code. GATE is an advanced opensource software developed by the international OpenGATE collaboration and dedicated to numerical simulations in medical imaging and radiotherapy [5].

In the simulations we have implemented the geometries of our experimental setups.

MBI system consists of a large detector which has the same dimension as a standard mammographic screen (150×200mm^2), and a smaller one with dimension allowing spot compression (70×70mm^2). Both detectors have been implemented in the simulation.

The first detector uses a 1 cm Tungsten parallel hole collimator with 1 mm holes and 0.34 mm septa. A 0.5 mm of Al is positioned right after the collimator coating a 6 mm layer of NaI, and a layer of 3 mm of glass coupled the scintillator to the photodetector.

The second detector has a 4 mm Tungsten pinhole collimator with a 2 mm hole. The NaI layer is placed 38 mm from the collimator. The image obtained with pinhole collimator has a certain magnification due to the detector geometry itself. The magnification factors have been calculated for different distances of the source from the collimator, and a processing of the simulated image was done taking into account those magnification factors to normalise the image in order to fuse it with the parallel hole one.

The detectors are placed in an opposite position one to each other and a 5 cm water phantom is placed between them.

GATE code permits to simulate the scintillation process of gammas from the source inside the scintillation layer. The optical photons generated are followed simulating their interaction processes inside the detector volume and their last position at the end of the glass layer is recorded.

An analysis code allow to study the simulation output and, taking into account the quantum efficiency of the photodetector, to reconstruct the position of the gamma from his optical photons generated and so to construct an image of the source.

The ^{99}Tc emission is simulated by a monochromatic source of 140 keV in energy.

For BSI we used the same detector as implemented in MBI changing some parameters of parallel hole collimator. In this case the collimator was a 20 mm Tungsten slab with 1.475 mm^2 hole dimension, and 0.305 mm septum dimension. An ^{90}Y source has been implemented.

3. Results

3.1. MBI

A monochromatic point source was used to determine the spatial resolution of the system. The following table shows the results of spatial resolution obtained for the two detectors for different distances of the source.

Table 1. Spatial resolution calculated from the simulations.

Source/Collimator distance	SR (mm) Parallel hole	Analytical values	SR (mm) Pinhole	Analytical values
1.0 cm	3.7	3.1	2.6	1.7
1.5 cm	4.0	3.7	2.8	2.0
2.0 cm	4.6	4.4	3.1	2.1
2.5 cm	5.1	5.0	3.3	2.2

The values of spatial resolution obtained are in good agreement with the analytical values calculated.

The intrinsic resolution of the crystal has been estimated using a directional source. The treatment at the interface has been taken into account. We determine the intrinsic resolution depending on the reflectivity or absorbant properties at the entrance and at the lateral walls of the crystal.

Every images was obtained simulating a background and an hotspot. The background source had the same dimension as the water phantom. The hotspot was a sphere with variable dimension in a range between 3 mm to 5 mm radius, and was placed in the middle of the phantom at different distances from the detectors: from 10 mm to 25 mm distance from the pinhole, and from 40 mm to 25 mm from parallel hole collimator respectively. The uptake used in this work was 10/1.

Images obtained with both detectors have been processed and a combined image was obtained by pixel to pixel values multiplication after proper normalisation.

Table 2. Intrinsic resolution of the crystal respect to inner wall coating of it.

Spatial resolution	Lateral walls	Entrance
1.15 mm	reflective	reflective
0.82 mm	absorbant	reflective
0.82 mm	70% absorbant	10% absorbant

The SNR (Signal to Noise Ratio) has been used to characterize the image. SNR values obtained are shown for different hotspot sizes.

Table 3. SNR estimated values for different hotspot radii.

Hotspot r = 5 mm

Dist. (mm)	Par	Pin	Combo
10/40	21	50	60
15/35	24	56	66
20/30	24	46	56
25/35	30	37	49

Hotspot r = 4 mm

Dist. (mm)	Par	Pin	Combo
10/40	10	36	37
15/35	13	38	42
20/30	15	31	35
25/35	19	27	31

Hotspot r = 3 mm

Dist. (mm)	Par	Pin	Combo
10/40	5	23	24
15/35	6	23	24
20/30	7	21	21
25/35	10	20	21

As is shown by the simulations, the use of a pinhole collimator allowing spot compression, which means that the detector can be positioned closer to the lesion, leads to a greater SNR values in the combined image and hopefully to a major detectability of smaller lesions.

3.2. BSI

We simulated a spherical beta-emitting source, 3 mm radius, positioned inside a water phantom. The Bremsstralung interaction positions were recorded to study the origin of the detected Bremsstrahlung gammas. In figure [1] a 3D map of Bremsstrahlung events distribution is shown as well as the events distribution along an axis. The events distribution results in a FWHM of about 9 mm for a spherical source of 6 mm in diameter.

As before, the image was reconstructed from the optical photons generated by the interaction of Bremsstrahlung photons with the scintillation material. In the upper part of figure [3] an example of reconstructed image is shown. A study in energy was performed, reconstructing images in different energy windows. At higher energy the directional information carried by the photon is compromised by the relative transparency of the collimators and the reduced photon absorption of the detector walls. On the other hand, low energy Bremsstrahlung photons still carry on direction information and actually provide clean, well contrasted, images.

The Monte Carlo results were preliminary compared to real data, obtained by two phantoms composed by Perspex cylinders with three small cylindrical cavities each, loaded with a $^{9}0Y$ solution. As shown in figure [2], one cavity is on the axis of the phantom, a second one is close the axis and a third cavity is near the phantom border.

The image acquired experimentally was studied as function of the energy. As for the simulation study, a very informative image resulted for lower energies as shown in the lower part of figure [3].

Journal of Physics: Conference Series **689** (2016) 012029 doi:10.1088/1742-6596/689/1/012029

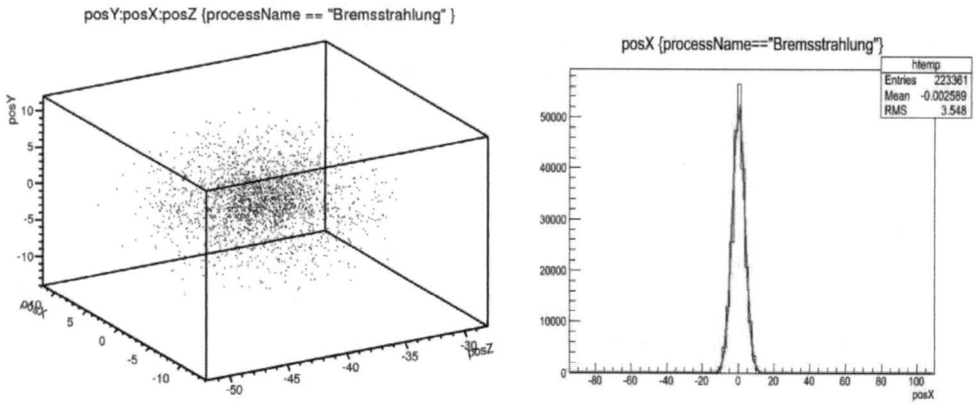

Figure 1. Left: Bremsstrahlung events distribution inside a water phantom. Right: Bremsstrahlung events distribution along an axis.

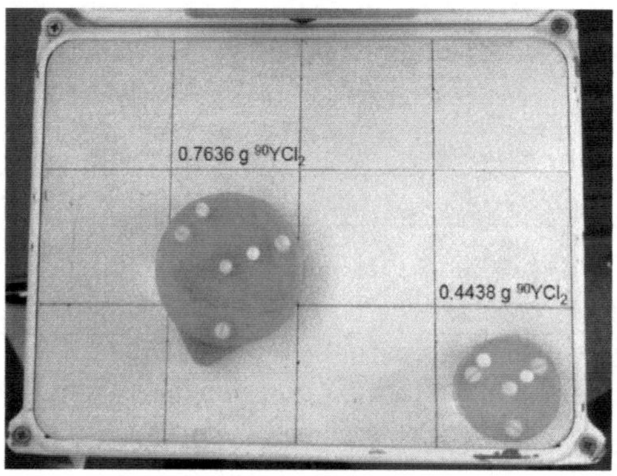

Figure 2. The two Perspex cylinders used in the experiments.

4. Conclusions

4.1. MBI

A good agreement between analytic calculation and simulated data as been obtained, somehow validating the Monte Carlo approach. The SNR of the properly combined image from the two detectors is noticeably better than the SNR of the single images; this will enhance the detection probability of small lesions, without affecting significantly the spatial resolution.

4.2. BSI

This preliminary study confirms the possibility to trace a beta emitter through Bremsstrahlung gammas. A compact gamma camera was adapted for Bremsstrahlung photon imaging and indirectly beta- distribution monitoring; real data have been acquired by a more complex system however results are in good qualitative agreement with the simulated Monte Carlo simplified model.

Further analysis work is necessary to complete the Monte Carlo studies, however the implemented model and procedures are flexible enough to permit the study of the response of a quite extended potential applications such as MBI and BSI.

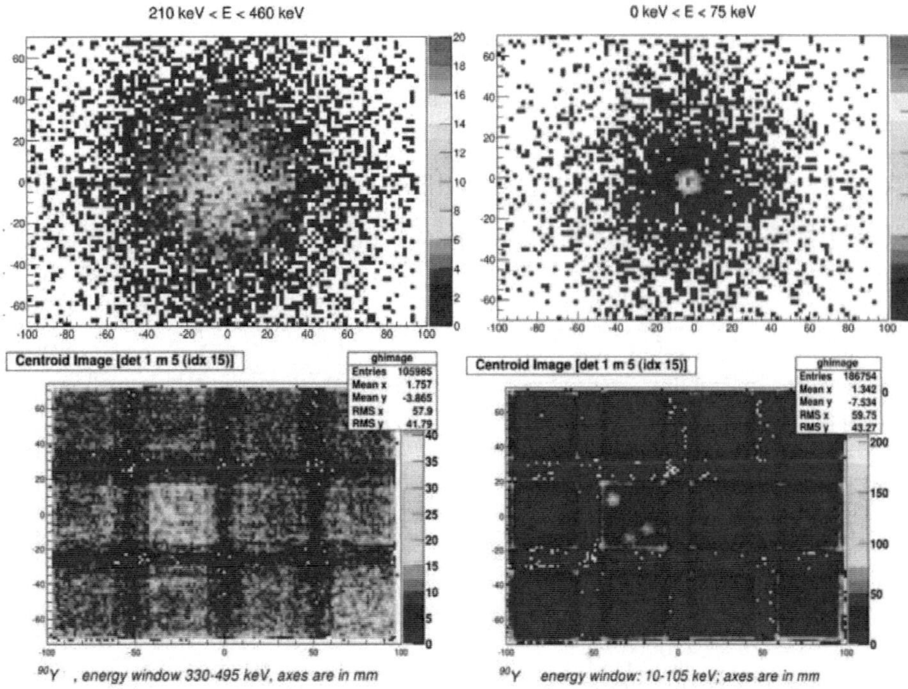

Figure 3. Top: simulated images from a spherical source 6 mm diameter, with different energy cuts: the lower energy photons provide the best image. Bottom: images of real phantoms with approximately the same energy cuts of the corresponding simulated images.

5. Acknowledgements

The authors thank Metaltronica for the financial support on the MBI work.
A special thank to Marco Capogni, without which the activities in BSI could not be carried out and that allowed us the access to the CRESCO cluster.
The computing resources and the related technical support used for this work have been provided by CRESCO/ENEAGRID High Performance Computing infrastructure and its staff [6]. CRESCO/ENEAGRID High Performance Computing infrastructure is funded by ENEA, the Italian National Agency for New Technologies, Energy and Sustainable Economic Development and by Italian and European research programmes, see http://www.cresco.enea.it/english for information.

References

[1] O'Connor M, Rhodes D, Hruska C 2009, *Expert Rev Anticancer Ther*, 9(8): 1073-80
[2] Hruska CB, Phillips SW, Whaley DH, Rhodes DJ, O'Connor MK 2008, *AJR Am J Roentgenol*, 191(6): 1805-15
[3] Cusanno F, et al. 2007, *Nuclear Instruments and Methods in Physics Research A*,**571**, Issues 1-2, 77-80
[4] Shen S, DeNardo L, Yuan A, DeNardo DA, DeNardo SJ, 1994, J Nucl Med,**35**, no. 8, 1381-89
[5] Morel C, et al., 2004, Phys Med Biol,**49**, Pages 4543-61
[6] G. Ponti et al., Proceedings of the 2014 International Conference on High Performance Computing and Simulation, HPCS 2014, art. no. 6903807, 1030-33

A Psychophysics experimental software to evaluate electrical pitch discrimination in Nucleus cochlear implanted patients

M T Pérez Zaballos[1], A Ramos de Miguel[2], M Killian[3] and A Ramos Macías[1]

[1] Departamento de Ciencias Clìnicas y Quirúrgicas, Universidad de Las Palmas de Gran Canaria, Las Palmas, Spain.
[2] Departamento de Aplicaciones Numéricas y Sistemas Inteligentes, Universidad de Las Palmas de Gran Canaria, Las Palmas, Spain
[3] Cochlear Technological Centre, Mechelen, Belgium.

E-mail: mtperezzaballos@gmail.com

Abstract. Multichannel electrode array design in cochlear implants has evolved into two major categories: straight and perimodiolar electrodes. When implanted, the former lies along the outer wall of the *scala tympani*, while the later are located closer to the *modiolus*, where the neural ends are. Therefore, a perimodiolar position of the electrode array could be expected to result in reduced stimulus thresholds and stimulating currents, increased dynamic range, and more localized stimulation of the neural elements. However, their advantage for pitch discrimination has not been conclusively stated. Therefore, in order to study electrode independence, a psychophysical software has been developed, making use of Nucleus Implant Communicator tools provided by Cochlear company under a research agreement. The application comprises a graphical interface to facilitate its use, since previous software has always required some type of computer language skills. It allows for customization of electrical pulse parameters, measurement of threshold and comfort levels, loudness balancing and alternative forced choice experiments to determine electrode discrimination in Nucleus© users.

1. Introduction

The level of hearing rehabilitation obtained through cochlear implants has increased dramatically since its introduction. This is the result of developments that include advanced signal processing, higher stimulation rates, greater numbers of channels, and the development of more efficient electrode arrays [1].

The results of several experiments in which electric field patterns for different electrode configurations have been investigated within the cochlea, reveal that the spatial gradient increases as the distance between the stimulating electrodes and neuronal tissue decreases. Therefore, an electrode that is closer to the neuronal endings would be expected to reach the hearing threshold current at lower levels due to the steeper slope of the electric field [2,3]. The result would be higher for sets of electrodes placed laterally. In addition, electric fields steeper slopes are less spread to adjacent neuronal populations, which decreases the probability of interaction between channels. This can result in a better discrimination between the electrodes [4-6].

However, the selectivity of the electric fields about the neural stimulation depends on a number of factors apart from the electrode-nerve proximity. These include: patterns of neuronal survival population, stochastic variability characteristics of individual nerve fibers, cochlear anatomy, signal characteristics and design features of the electrode [7]. Many of these factors may be interdependent and may affect the threshold as well as the selectivity. For example, an individual with a poor nerve survival will have poorer selectivity and higher thresholds than one with good nerve survival, regardless of the type of electrode array, as the subject requires higher values to achieve optimal neural excitation.

Several research groups have designed psychophysical platforms to reduce the amount of programming needed to build experiments. Although these are very comprehensive and adaptable tools that allow the researcher to design virtually any experiment on psychoacoustics, they lack the necessary tools to perform electrical stimulation experiments [8-11].

An exception to the limitations of these platforms is APEX, a software application built and distributed under contract by the Experimental Audiology Department in Leuven [12]. This platform supports psychoacoustic experiment design both for acoustical and electrical stimuli by means of a cochlear implant, and even the combination of both. In APEX, the actual experiments are designed in XML format. Even though this is a great advantage and requires much less learning time, it still requires some degree of computer language knowledge, since XML is not as straightforward as a visual interface.

This paper describes a visual interface-based psychophysical software to conduct electrode discrimination tests to work towards better understanding of the psychophysics of electrical hearing. The platform is controlled exclusively through a graphical interface, where pulse parameters are entered in purpose-designed cells and loudness balance and alternative-forced-choice experiments are already built in.

2. Platform design

The present Psycoachoustic Research Platform was designed using the Nucleus Implant Communicator (NIC) library for Python (Python Software Foundation, v2.3). NIC is a research tool developed by Cochlear LTD and allows researchers to build applications to control the electrical stimuli delivered by the intracochlear electrodes of Nucleus Cochlear implants.

The researcher and subject interfaces were designed using Visual Studio (Microsoft Corp. Visual Studio Community 2013). It was also used to collect information about the connection state of the implant. Since the NIC library is in Python, scripts in the same language were built to control the implant receiver/transmitter by sending instructions to a supplied cochlear implant research sound processor. The supplied processor was connected to a computer via a USB port. Thus, a file-based communication system was established. Finally, the data is arranged so that a patient can be assigned to a range of psychoacoustic tests nested within the type of stimulus and further dependent on intensity. For each patient, the user can define multiple tests.

3. Platform Description

Insert Patient

The option *Insert Patient* on the Start page opens the window in Fig. 1, where demographic and clinical data can be inserted. The Comfort (C) and Threshold (T) level values from the patient's standard map can also be inserted. Finally, the location of the electrodes with respect to the modiolus can also be inserted as perimodiolar, mid-modiolar and lateral wall.

On this window, and all subsequent ones, there is an indicator of the communication state between the application and the processor (connected/not connected) for safety reasons. It will show whether the computer correctly detects the processor or not.

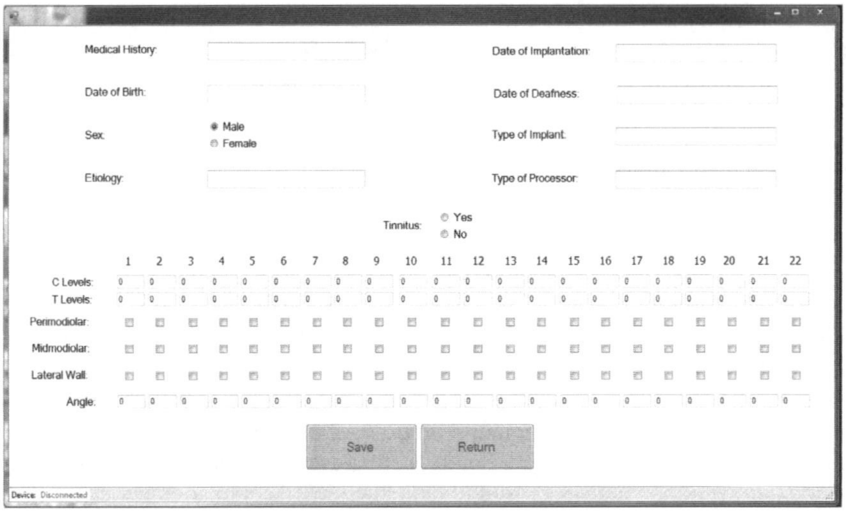

Figure 1. Insert Patient. Here demographic, clinical data and electrode position with respect to the *modiolus* can be registered.

Stimulus Definition

Once the patient has been saved, the Existent Patient window replaces the New Patient, where the same information is displayed, but now the user can select the different tests to perform the study.

A stimulus must be defined before the experiment can proceed. The Define Stimulus button opens a window (Fig. 2) that allows customization of the following parameters: stimulation mode, stimulus duration, inter-stimulus interval, pulse phase and gap durations and pulse frequency. The stimulus consists of a train of electrical biphasic pulses at a frequency specified by the pulse rate option. Stimulation mode refers to the type of ground electrode used: the reference electrode located in the mastoid region (1), the internal processor case (2) or a combination of the two (1+2). The inter-stimulus interval refers to the resting time between two stimuli. These settings will remain fixed for all tests of the study.

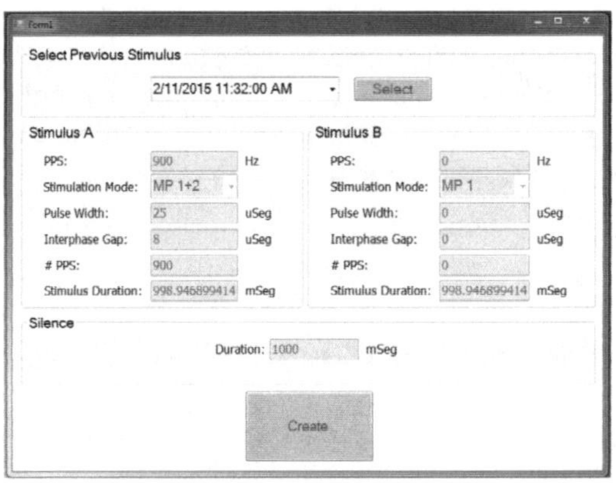

Figure 2 Stimulus definition window. PPS: pulses per second; Stimulation mode: ground electrode location (1 for ground electrode in the mastoid, 2 for internal processor case and 1+2 for a combination of both); # PPS: number of pulses per second (defines stimulus duration); silence: defines time lapse between pulse trains.

Dynamic Range Map

The Dynamic Range (DR) of an electrode defines the current intensity span that produces auditory sensation and is determined by the T and C levels of each electrode. The window can be seen in Fig. 3. The T level is calculated using the *up-down* method: a stimulus is presented at progressively increasing intensity until the patient reports hearing sensation and then it is lowered down again using smaller intensity steps until it is no longer heard. To find the C level, the stimulus is presented in progressively smaller steps as the intensity rises until the patient reports that the sound is no longer comfortable because it is too loud, but never painful.

In this part of the test, a new map can be created or a previous session can be loaded by selecting one from the *Load Session* scroll. When two different stimuli are used in an experiment, two independent maps must be created using one stimulus at a time.

Loudness Balance

The second step is to loudness balance all electrodes so that the stimuli delivered to all electrodes are perceived as equally loud. In the *Loudness Balance* window shown in Fig. 4, the user can compare an electrode with 21 other electrodes. In addition, one can select the stimulus to be used (A or B). The user must select the percentage of DR intensity that the initial reference electrode will have. This will be the electrode that will be objectively fixed at the desired percentage, and all others will be perceptually balanced with respect to it. The number of times one electrode will be compared to another (repetitions) can also be selected.

The method used is the confluence method. The patient is presented with two stimuli, the first corresponding to a reference electrode, and the second to a test electrode. The latter begins randomly above or below 10% of the DR percentage selected prior to test commencement. The patient selects which one of the two stimuli sounds louder using the window in Fig. 5 and the intensity of the test electrode will be lowered or raised until confluence is achieved. Then the patient has to state that they sound equal using the appropriate button. The stimuli will be repeated with the same settings to confirm that the patient perceives them as equal and then the system will automatically start again but with the test electrode being 10% different in the opposite sense. The test electrode final value is the average of the two approaches.

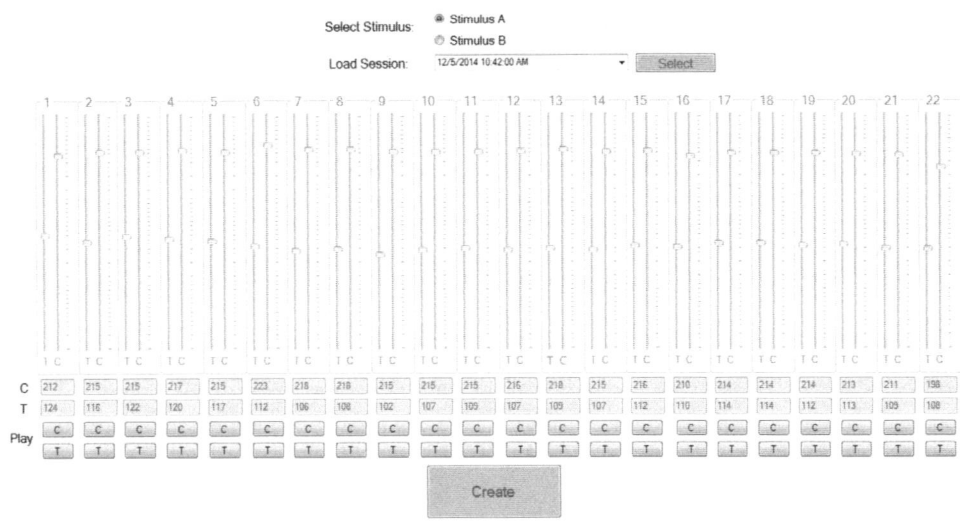

Figure 3. Dynamic Range Map. C and T levels for each electrode are determined using the up-down method.

Journal of Physics: Conference Series **689** (2016) 012030 doi:10.1088/1742-6596/689/1/012030

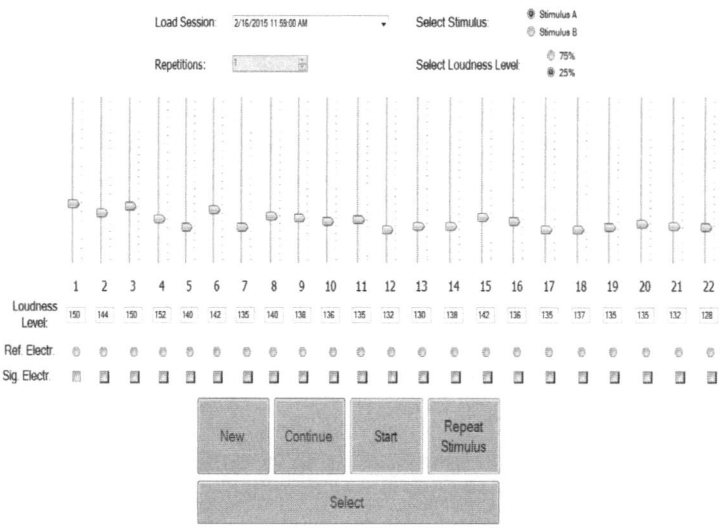

Figure 4. Patient view for electrode loudness balance. The patient has to select which stimulus sounds louder.

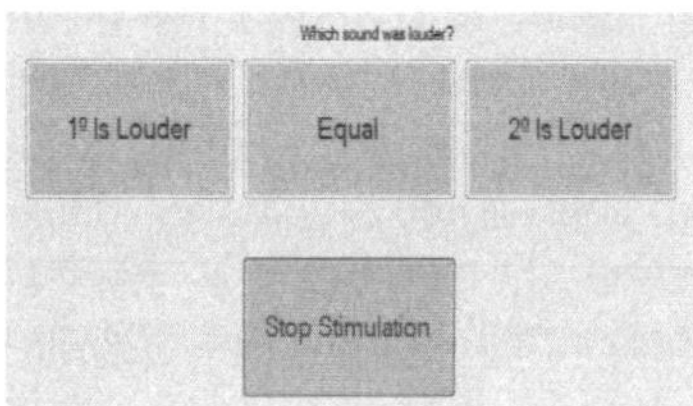

Figure 5. Loudness Balance Window.

Alternative Forced Choice Experiment for Electrode Discrimination

The chosen method for electrode discrimination is a three-interval, forced-choice procedure (3FC) (Fig. 6). Two of the stimuli come from a selected reference electrode and a third one from a signal electrode. They are presented in random order and each test electrode is presented the number of times stated in the *Number of Repetitions* scroll. The patient has a user interface with three buttons (stimuli 1 to 3) and has to select the one that sounds different. The researcher will select a reference electrode and a number of electrodes to compare it with. Results are given as percentage correct scores.

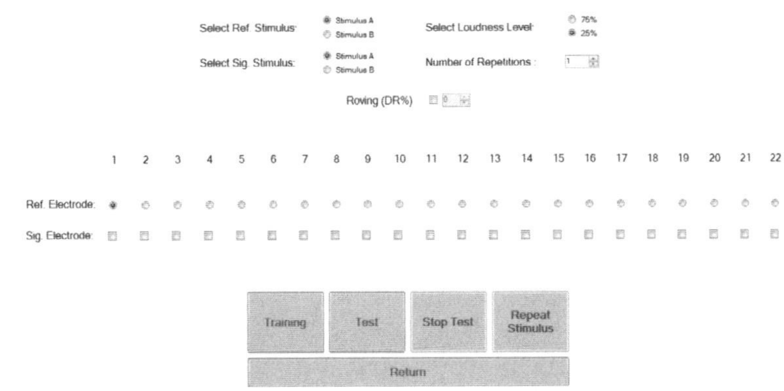

Figure 6. Alternative forced choice experiment for electrode discrimination.

4. Conclusion

The experimental software developed in this paper allows for conducting psychoacoustic experiments on electrode discrimination. The stimulus variables and psychophysical procedures can be adjusted and the three basic steps necessary to conduct electrical stimulation experiments are easy to implement. The side benefits are to relieve psychoacoustic researchers of the burden of programming language skills, which would otherwise be necessary to generate and analyse stimulus signals. However, the application is experimental and can only be used by professionals who are trained on the application and have signed a NIC agreement. A NIC agreement requires a signed contract with the company to allow the researcher to access the Python library that can control the processor.

References

[1] Wilson BS, Dorman MF 2008 *Hear. Res.* **242**(1):3-21.
[2] Rattay F, Lutter P, Felix H. 2001 *Hear. Res.* **153**(1):43-63.
[3] Briaire JJ, Frijns JHM 2000 *Hear. Res.* **148**(1):18-30.
[4] Frijns JHM, de Snoo SL, Ten Kate JH 1996 *Hear. Res.* **95**(1-2):33-48.
[5] McKay CM, O'Brien A, James CJ 1999 *Hear. Res.* **136**(1):159-164.
[6] Pfingst BE, Holloway LA, Zwolan TA, Collins LM 1999 *Hear. Res.* **134**(1):104-115.
[7] Forster KI, Forster JC 2003 *Behav. Res. Methods Instrum. Comput.* **35**(1):116-124.
[8] Lopez-Bascuas LE, Marín CC 1999 *Behav. Res. Methods Instrum. Comput.* **31**(2):334-340.
[9] Hillenbrand JM, Gayvert RT 2005 *J. Speech Lang. Hear. Res.* **48**(1):45-60.
[10] Kwon BJ 2012 *Behav. Res. Meth.* **44**(2):361-373.
[11] Geurts L, Wouters J 2000 *J. Acoust. Soc. Am.* **108**(6):2949-2956.